Frederic Carpenter Skey

Hysteria : remote causes of disease in general, treatment of disease by tonic agency,

Local or surgical forms of hysteria, etc.: six lectures, delivered to the students of St. Bartholomew's hospital, 1866

Frederic Carpenter Skey

Hysteria : remote causes of disease in general, treatment of disease by tonic agency,
Local or surgical forms of hysteria, etc.: six lectures, delivered to the students of St. Bartholomew's hospital, 1866

ISBN/EAN: 9783337689643

Printed in Europe, USA, Canada, Australia, Japan

Cover: Foto ©berggeist007 / pixelio.de

More available books at **www.hansebooks.com**

HYSTERIA.

Remote Causes of Disease in General. Treatment of Disease by Tonic Agency. Local or Surgical Forms of Hysteria, etc.

SIX LECTURES,

DELIVERED

TO THE STUDENTS OF ST. BARTHOLOMEW'S HOSPITAL, 1866.

BY

F. C. SKEY, F.R.S.

SECOND AMERICAN EDITION.

New York:
MOORHEAD, SIMPSON & BOND.

1868.

NEW YORK:

AGATHYNIAN PRESS, 60 DUANE STREET.

TO

THE MEMORY OF

Sir Benjamin Collins Brodie, Bart., D.C.L.,

"THE GREATEST OF MODERN SURGEONS."

PREFACE.

THE following Lectures were delivered to the Students of St. Bartholomew's Hospital, and they are now published at their request and for their instruction.

If any members of older growth fail in knowledge of matters so essential as those I have discussed, however unskilfully I may have treated them, the sooner they enter on the path of inquiry the better.

It is in the nature of scientific inquiry to intermingle error with the truth. Our faculties, however cultivated, cannot penetrate the recesses in which are concealed the more precious gems of perfect knowledge, the value and even the nature of which are variously appreciated by different observers. Hence opposite deductions and contradictory opinions.

The majority of our profession are biassed in favor of depletive measures. They consider diseases to originate in excess of vitality, and to be the product of undue force; I have endeavored to show that they far more commonly arise from reduced or exhausted power. A weak condition of the animal body is intelligible enough, but an abnormal condition warranting a reduction of vital power by artificial agency I cannot understand.

We talk of "increased heat," of "increased action," of "excitement," and so on, as demanding a reduction or a lowering of the vital powers; while it is notorious that intense heat of skin is the frequent attendant on the last hours of life, and who will venture to treat the "excitement" of *delirium tremens* by any agent less potent than alcohol or opium?

The great machine of Medical knowledge, however, is moving onward in the direction of truth.

On a line between the antagonistic principles of the schools of Cullen and of Brown the Profession will eventually take its stand—at a point, I suspect, far nearer to Brown than to Cullen. This is the natural reaction following on years of error.

The palpable confusion between diseases of the vascular and of the nervous systems which has so long prevailed is but evidence of the same error.

Whether the sketch I have drawn be overcharged—for it does not affect the pretension of a finished picture—I leave to be determined by others.

CONTENTS.

LECTURE I.
PAGE

Congestion not Inflammation—Treatment of congestion—Cases of congestion—Tonic treatment exemplified in abscess—Evils of depletive treatment—Tonic treatment—Bark, wine, &c., the quantity determined by the pulse—Employment of wine in the London Hospitals.   9

LECTURE II.

Remote causes of Disease—Causes definite and traceable—General cause of disease—The aperient system—Treatment by increasing the quantity of healthy blood—Stimulants in debility reduce the frequency of the pulse—Alcoholic stimulants essential—Cases—Diseases of animal and organic life—Cases in illustration of the two varieties of disease.   26

LECTURE III.

Imitations of real disease—The study of, much neglected by the profession—Indicated by local pain and muscular spasm—Distinct from Inflammatory disease—The term "Hysteria" objectionable—Nature and source of Hysteria—Errors in diagnosis—Critical examination of cases—Hysteric affection of joints—Hysteric affections of the spine—Hysteric contraction of fingers. 43

LECTURE IV.

Symptoms of common paroxysmal Hysteria—Constitutional liability—Nerve and nervous system—Effect of railway accidents—influence of the mind—Effects of an unstrung nervous system on the actions of daily life—Imitative or contagious Hysteria—Surgical cases of—Distinct from Neuralgia—Relation between Hysteria and the brain or spinal cord—Hysteria combined with real disease 59

LECTURE V.

Distinctions between nervous and vascular diseases recapitulated—General localities of Hysteria—Case of Hysteria of the muscles of the larynx—Hysteric affection of the mammary gland—Value of exercise—Relative value of foot and horse exercise—General and local treatment—Hysteria of hypochondriac regions—Spinal affections—Cases—Efficacy or inefficacy of issues—Cases—Railway actions and extortions—Cases—Hysteric joints, treatment of—Cases 73

LECTURE VI.

Hysteric affection of the Œsophagus—Hysteric affection of the stomach; —Gastrodynia—of the ovary—Hysteric contraction of muscles—Wry neck—Contraction of fingers—Contraction of elbow-joint—Hysteric contraction of the muscles of the leg and foot—Hysteric paraplegia—Hemiplegia 94

SIX LECTURES ON HYSTERIA.

ON THE REMOTE CAUSES OF DISEASE, AND ON ITS TREATMENT BY TONIC AGENCY.

FIRST LECTURE.

Congestion not inflammation—Treatment of congestion—Cases of congestion—Tonic treatment exemplified in abscess—Evils of depletive treatment—Tonic treatment—Bark, wine, etc., the quantity determined by the pulse—Employment of wine in the London Hospital.

GENTLEMEN : Circumstances have limited my opportunities of instruction. I purpose, in the few lectures I shall deliver, to discuss such principles of Medicine as may prove useful to you, which involve the essence, if I may so term it, of Medical and Surgical practice, and the assent to and adoption of which appear to my judgment indispensable to success in the treatment of disease. If these principles are based on error, you will readily detect it, and they will present themselves as so many rocks and shoals in your future career to be avoided as dangerous. If sound, you will approve and adopt them. I adopt them because I am satisfied, theoretically, they are based on sound views of the animal economy ; and practically I find them in a pre-eminent degree more successful in the treatment of disease than those which refer so large a proportion of disease to what it termed *inflammation* and its supposed consequences.

It is a prevailing error among the unthinking members of our profession which refers all examples of redness of vessels to inflammation. No term can be more inappropriate. The ancient definition of inflammation is founded on truth, and will hold true to the end of time; but it is rendered absurd by the practice of the modern school. In the too frequent employment of the name we forget the conditions essential to it. The arterial and capillary systems perform their healthy functions so long only as they retain the normal influence of the nerves supplying them, whether locally or generally. If the supply of nervous influence fails, the capillary system of a part loses its healthy tone, and the vessels locally dilate. At the same time they lose the power of propulsion, and there arises of necessity a local *remora* or arrest of the circulation, and, as a matter of course, the part becomes red. In this inflammation? Yes, in the judgment of many, it is termed a form or variety of inflammation, not strict or positive, but inflammation of an asthenic type, although it fails in three of the four conditions essential to true inflammation, and exhibits redness only, wanting pain, heat, and swelling.

Now these examples of local congestion or plethora of vessels are yet in the present day confounded with true inflammation, and are treated by what we term antiphlogistic agents—namely, general and local bleeding, purgatives, diaphoretics, and the suspension of food. True inflammation is not so commonly seen as you may imagine; and inasmuch as local redness does not involve all the required signs of heat, redness, pain, and swelling, which that state requires, it cannot be true inflammation, and there is no difficulty in proving that the treatment usually adopted in true inflammatory disease is especially objectionable and injurious in this.

There is another variety of so-called inflammation, in the mouths of a large number of us called *chronic;* what it means is somewhat difficult to define, but we employ it rather to conceal a difficulty, than to expound it.

There are two principles on which this congested condition of the blood-vessels may be treated;

1. Blood may be taken locally, the system may be lowered by salines, as they are termed, by purgatives and diaphoretics. This treatment by local depletion would lead to the inference that the blood is at fault and should be removed.

2. The blood in these vessels may be forced onward by giving increased action to the heart, by the resort to agents that tend to restore the healthy tonic condition of, or, at all events, to give force to, the capillary system of the affected part.

This latter is termed the tonic treatment of disease: a principle, I consider, based on a sound view of the functions of the animal economy, and which formed the staple of my treatment during the period in which I was attached to this Hospital as Surgeon, and for many years prior. If this principle be sound, treatment by depletion in any form must be an error, and therefore injurious, and that it is injurious I have not a doubt. I can speak with some confidence, for

I have tried both. In the history of the human mind there is no operation more difficult than that of divesting it of early impressions inculcated by authority and confirmed and established by time. Convictions increase in strength as we get older, for the habit of examining the one side of a question only, and of conforming to all its requirements, has assumed a settled, permanent form, and with the multitude engaged in the practice of our profession there is no adequate motive to enter on a path of inquiry which may tend to unsettle the convictions of their past lives. Some fifteen years since I had a succession of cases of severe congestion—that is, redness, not inflammation—accompanying wounds caused by operations, cases of mammary cancer or other tumors removed by the knife, amputations, hernia, etc. These cases exhibited on the third or fourth day congestion of the vessels around the wound, which often extended and led to abcesses, and occasionally to death. It was called in common hospital parlance erysipelas, having the feature of redness only in common with that disease; but it was not erysipelas: in truth, it was a far more formidable condition. I ordered every patient, after undergoing any serious operation, treatment by stimulants. I gave two tablespoonfuls of brandy and two of water every four hours, to be continued for one, two, or more days. From that date I never lost a patient from this cause, and rarely did any unhealthy condition of the affected region present itself.

Now observe this feature in the treatment by stimulants. If these examples of local congestion of the blood-vessels partook of the nature of inflammation, you, I am sure, will concur with me in considering this treatment a blunder. If stimulants be inadmissible in the treatment of inflammation, the treatment should have proved injurious. This it did not prove, but, on the contrary, effected early and curative results; therefore the disease was not that of inflammation. What is true of the local is equally true of the general.

The late Mr. Jones, of Jersey, was a remarkably successful operator, and in reference to his cases of operation for the excision of joints (considered to be diseased) he operated on a succession of more than twenty-five without the loss of a single case, while in England at the same date the mortality was always great, recovery being rather the exception, and not the rule. What was the secret of this? Was it fresher air, or different management? I made the inquiry, and learned that Mr. Jones invariably gave every patient on whom be operated at least a pint of port wine on each of the two days following the operation, and he acknowledged to me that he owed the treatment which had been attended with such remarkable results, to his observation of the success that had attended my treatment in St. Bartholomew's Hospital. The vital powers of the system are always depressed by a large operation, whether much blood has been lost or not. I have found this combined tonic and stimulating treatment eminently successful, and so convinced am I of its value that, if I were told it had been less successful in the hands of others, I should feel assured it had never been fairly tried.

In no disease is the application of this principle more valuable than in every variety of abscess, whether chronic or acute, and especially perhaps the latter. Let us take a common example. In two or three weeks after her confinement a lady has a milk abscess. You are called to attend her, and on examination you find a firm lump in the substance of the breast, involving perhaps one third of its circumference. It is painful and heavy. The diagnosis is clear enough; it is the deposit of an incipient abscess. What is your treatment? I will tell you. You think you can disperse it. Impossible. You order two or three grains of calomel with some colocynth or other purgative; you follow this up with salines, a class of drugs that I consider more than any other displays the vacillation and uncertainty of the practitioner; and you order leeches, about eight or ten in number, and warm fomentations and poultices, to be applied to the breast. All these remedies are resorted to under the impression that you can obtain absorption of this solid mass. But the attempt fails: it always has failed; it always must fail; and yet it is nearly always attempted. Doubtless the leeches relieve the pain of distension, and that is all; but this good does not counterbalance the injurious effects consequent on the loss of blood required in the body for the most important of purposes. At the Examination Board of the College of Surgeons, this question as to the treatment of the early stage of a mammary abscess has been for years a critical question of mine. The reply has been invariably "leeches, fomentations, poultices, purgatives, salines." Does this reply indicate more of knowledge or of ignorance? Such is our profession among the multitude, deny it who can.

Now, before I tell you what is right to be done; let me make a remark or two in passing. This lady's condition is marked by great debility. You will find on inquiry either that she had a "hard time," as it is called, that she lost blood largely during the birth of her child, or that subsequently to her confinement she has been the subject of some debilitating cause, such as diarrhœa, loss of food, or fatigue of body or mind, or some other depressing influence has drawn largely on her physical strength. It is owing to this depressing influence that she is the subject of this painful disease. *Had the weakness not existed, she would have escaped milk abscess.*

You do not find milk abscess or any other form of abscess in a person of vigorous circulation. It is the product of weakness, whether inherent in the constitution or the result of accident. For evidence of the truth of this fact, examine her pulse. Is it firm? Will it bear the pressure of the finger without stopping it? Is it moderate in the number of its beats? Is it at or under 70? No; it is at 80, or 90, or more; it is weak, and easily compressed by the finger; it may be large, and full, and rounded, or it may be small; but be it one or the other, it is stopped by slight additional pressure of the finger. Either the blood is deficient in quantity, or the heart fails in power: probably both. And in this condition of the system, with indications of debility perfectly palpable before you, you appeal to the great British panacea for four fifths of all diseases, and you prescribe for your patient calomel and colocynth, leeches, salines, and fomentations! Then, in my capacity of examiner, I generally proceed thus—Very well; and now tell me on what principle and for what purpose you administer these agents. Let us begin with the calomel. What object have you in view? To regulate the secretions. What secretions? The liver. What has the secretion of the liver to do with an abscess of the breast? Do you necessarily infer disordered liver because the mammary gland is the seat of abscess? Can any opinion be more irrational?

I often marvel at the wonderful faith exhibited by medical men in the influence of these " secretions," as though the liver, the most inoffensive organ in the whole body, was the centre or focus of all its maladies. Well! the calomel and the other depressing agents do their work, and what are the consequences? The patient is weaker; whatever nourishment remained in the body is taken from it. The supply of food is arrested. Salines follow, and draw on the circuiation yet more positively, and there remains the tumor as at first. The pulse is now 90 or 100, weaker, and yet more easily compressed. The appetite for food is gone, and in order to obtain sleep, which is incompatible with her present condition of weakness, you are compelled to resort to sedatives at night. Against these troubles which have interrupted the natural progress of the malady the constitution struggles, and in due course of time the mass suppurates, but not healthily. It softens in the centre only, and *points*, and you puncture it. The quantity of matter that escapes is small in proportion to the solid mass it leaves behind, and weeks elapse before this patient is restored to health.

Such is briefly the history of an ordinary case of mammary abscess; the practice adopted is general, though happily for humanity not invariable, and I protest, as a Hospital Surgeon whose opportunities of observation have not been inconsiderable, against such principles of treatment, as irrational and preposterous, which arrest the functions of Nature in her attempt to bring this disease to an early and curative crisis, by the adoption of a false and radically unsound doctrine of the old school of medicine, that all diseases partake of the character of inflammation, and are controlled by depletive agents.

Now let us take another view of this subject; examine the evidence closely and think for yourselves. Your first object is to look for the *cause* of disease. If you find by inquiry and observation that mammary abscess prevails only in weakly persons, as I have stated, that the subjects have been ill fed or are in impaired health, that they have a weak pulse and bad appetite for food, what treatment would you select, supposing that in such a weakly condition of the health no abscess existed? Would you not adopt a tonic treatment? Would you not administer wine and bark and nutritious food, even though, true to your early impressions, you threw in an occasional aperient under the title of an alterative? Well, then, can the presence of incipient abscess of the breast—itself a mere symptom with many others, and which owes its existence to debility and debility alone as the primary cause—can the presence of mere abscess justify you in the adoption of remedies which tend to increase the very evil you are endeavoring to counteract? You want to give strength, and you add to the weakness. You would not prescribe calomel without the abscess; why do you prescribe it with abscess? You may reply that you hope to carry off the abscess by purgation. But this is a delusion. What ground have you for hoping to carry off the abscess by purgation? Have you observed one single example in the course of years in which the employment of calomel, etc., has cured, or even reduced in size, an abscess of the breast? No! You have not seen it, nor will you see it until Nature abandons her own laws to oblige you. Now, gentlemen, if you could see this picture as I have sketched it, you would shrink from the folly that prevails so largely in the profession on this matter. You may argue in justification that you have attended lectures on the practice of surgery, and that you have been taught it. Very probably you have; but who taught the lecturers? Believe me, the sooner you unlearn these errors the better both for your patients and your own credit, and for the honor of our profession.

Presuming these principles to be unsound, however generally adopted, for that is no conclusive evidence in their favor, let us now consider the subject from an opposite point of view.

I assume a state of debility at starting. If you doubt my title to do so, observe and inquire. Inquire of men largely engaged in the practice of midwifery. Are strong and healthy, or are weakly women more liable to mammary abscess during lactation? You will receive but one reply. If, then, weakness prevails in the system so universally before the occurrence of the abscess, there is no escaping the inference that the weakness is the *cause* of the abscess, for you do not meet with cases of abscess in strong healthy women, but it selects for its victims the poor, ill fed, and impoverished. One of the worst cases I ever saw was that of a young woman in St. Bartholomew's Hospital, whose husband struggled to support a family of seven persons on about as many shillings a week. This woman was confined with twins, which she half nourished on a daily allowance of half a pint of bad beer, and for which misery the funds of St. Bartholomew's Hospital did ample penance.

You see, then, distinctly enough, both *cause* and *effect*. The cause is debility, the effect is abscess, commencing in a deposit of a large mass of coagulated fibrin. Now, which do you select for treatment, the cause of the disease, or the effect? If the effect, you leave the cause untouched for the production of another abscess. It is quite true that under some exceptional circumstances, by reason of an urgent necessity, we are compelled to direct our remedial powers against effects which imperil important structures, and even life itself; but that necessity does not hold in the case of a mammary abscess, which is an evil on a smaller scale. Should we not consider the abscess as a mere symptom of debility in common with other symptoms, such as a weak pulse, a hot skin, loss of appetite and sleep, constipated bowels, headache, if it exist, etc. ? And would you for a moment think of treating these symptoms individually? No! Then why thus treat the abscess, itself a mere symptom, and most especially when such treatment, itself essentially and radically false to the principles of sound pathology, drives you to the resort to remedies which, while they aggravate the evil of the cause by reducing the strength to a yet lower level, leave the effect untouched? Is this wisdom or folly?

Now let me ask this important question—Can any evil result to the constitution from leaving the abscess alone? I mean, as a local affection, can it do harm? I do not refer to it as a source of constitutional irritation—that is an evil which calls for aid—but is it probable that the health can suffer from its absorption, supposing such a local change possible? Assuredly not. Then it is obvious that the local affection, which cannot be got rid of by the means suggested, or by any other means that the ingenuity of man has yet discovered, should be let alone, and that the influence of the curative art should be directed to the constitutional health as the primary cause of the derangement.

And here, in all humility, and in the candid acknowledgment of our impotence to deal with disease, excepting through the hand of Nature, let us look to the real philosophy of our art, and throw ourselves on her ample resources for such aid as man cannot command, but which Nature is always willing to afford. Let us endeavor to restore the constitution to that condition of health and vigor it possessed before the crisis occurred which led to all this mischief, and, depend upon it, in obedience to a great law of Nature, *which takes cognizance of all disease with a view to remove it when removable,* you will find your tumor gradually disappear, or, what is more probable, it will pass quickly through those local changes essential to recovery, whether by gradual absorption, or by rapidly formed but limited suppuration, the one or the other alternative depending on the early or late resort to rational treatment.

This treatment is essentially tonic in its character. Its object, as I have above stated, is the employment of such medicinal and dietetic agents as tend to restore the constitution to a condition of health and vigor. This is all we can pretend to do, and, rely upon it, Nature will do the rest. I do not say, exclude all local treatment. Doubtless a few leeches may relieve the pain, and fomentations or poultices may in some slight degree contribute to the same end; but these agents are not, critically speaking, curative; they palliate symptoms only; and, as pain is an evil, and itself, when severe, a source of prostration, the surgeon may justly combat it by such remedies as draw as lightly as possible on the circulation and do not reduce the vital powers of the body.

There are two agents, in addition to good nourishment, especially applicable to the treatment of a case of this kind —bark and wine. There is an inferior class of remedies, but of the same character, such as bitter infusions, cascarilla, gentian, and the like, and the varieties of infusions of malt. There is no comparison between the virtues of the two classes, but as poor persons cannot afford wine, they may afford good beer, of which the strongest is the most efficient; but among medicaments *cinchona bark* has no rival. In all cases of abscess bark is invaluable. It promotes suppuration when suppuration is inevitable, and it checks suppuration in the exhausting action of chronic abscess. It promotes appetite, and gives vigor to the system beyond any other known tonic, and it is only on this principle that it carries our patients rapidly through the different stages of disease. I prefer the simple tincture of bark, and I recommend you to administer it in full doses of from one to three drachms twice a day. For adults I generally recommend a dessert-spoonful, and not less, and I have not observed that its employment in such doses demands the corrective influence of either purgative or alterative medicine; but, on the contrary, as I refer any temporary torpidity of the actions of the alimentary canal to the want of that vigor of system, and to that only, which the bark is capable of correcting, so, as a rule, subject of course to occasional exceptions, I leave the bowels alone. I have a very limited faith in the cathartic system. In this I know I am un-English; but I am content to live under the obloquy. On the subject of wine I have no intention of entering on the question how far alcohol is a source of strength or of nutrition. As an ingredient in the diet scale of the invalid, it is indispensable; and there is a very important rule of conduct as regards its administration, the knowledge of which I am most anxious to impress on your minds, and which involves a principle that cannot fail to obtain your concurrence on a closer observation of the nature of disease and the remedies pertinent to it. The prominent rule I allude to is this: Administer wine or alcohol in quantity regulated by its effects on the system, and not in any fixed quantity determined by usage. A gentleman with a weak pulse is told by his physician to take wine freely, and he indulges in a bottle of claret at and after dinner, and by so doing he considers himself to have obeyed the injunction of his medical adviser. Now I shall select port wine as the standard by which I regulate the advice I give under the above circumstances; and I am of opinion that for the purposes of health three or four glasses of wine is the maximum quantity that, *taken at any one time*, can be serviceable. All beyond this, answers the purpose of luxury and nothing more, and is more or less injurious. Suppose a person of abstemious habits takes a glass of wine, being much fatigued from exercise, how long will he continue to feel the benefit of it? Possibly, at the outside, two hours or less. The wine did good service at the time; but the effect has passed entirely away. If this person be subject to a renewal of the fatigue, he can take a second glass at the expiration of two hours on the same conditions, and so on. He has in his fatigue the equivalent to disease, and he takes the restorative, with this result, that, having consumed five or six glasses of wine, he is not conscious, judging from his sensations, that he has taken any.

If this law be founded in truth, I say, administer wine frequently in moderate quantities, or in immoderate quantities *if they are required*. You can judge for your patient what quantity he can take and what he requires, and I say unhesitatingly, when required, the more he takes the better and the sooner he will be well, which, I presume, is the object to be attained. Judge the quantity of wine then required by its effects, and do not administer it in given quantities.

I consider wine indispensable in the tonic treatment of disease. It should be administered in moderate quantities at short intervals. If too short the effect of the wine will prove cumulative, and reaction will follow. The art of administering alcohol, whether in the form of wine or brandy, is learned by observation of its effects on each individual; and be assured of this, that the quantity consumed with impunity, be it large or small, *is the gauge of its want by the system of that person.*

During the last forty years the treatment of disease has undergone great and important changes, and up to a later date the use of wine was very exceptional. I do not overstate the case when I affirm that within the period I have mentioned, the consumption of wine and brandy in the London Hospital has increased at least fourfold; and I may here relate an anecdote which confirms the above statement. It has, I believe, been before published, but I deem it right to bring it again under your notice. In the year 1848 the Treasurer of the Hospital commented on the quantity of wine I ordered for my patients. He said the Hospital could not warrant the large expense. I inquired the number of patients—550; and the wine consumed was three pipes per annum. I told the worthy Treasurer that the consumption of wine quite surprised me; that I could not understand how my colleagues could manage their cases, for that I could not treat Hospital cases without wine; and I assured the Treasurer I would do my best, for the credit of the Hospital, to raise the consumption of port wine from three pipes a year to thirteen, and that nothing less would satisfy me or my convictions. Twelve years elapsed, when I was again addressed by the successor of the then Treasurer on some matters connected with the diet scale of the Hospital. On inquiry into the consumption of port wine, he appealed to the Apothecary, who, referring to his wine-book, announced the quantity annually consumed to be thirteen pipes! I was on the previous occasion the great delinquent, but now, as the Treasurer declared, "You are all nearly equally bad, although *you* still head the list." And the same change in the treatment prevails more or less in every Hospital in London. Although this employment of stimulants and tonics at the expense of depletive agents has become so prominent a feature in modern practice, it does not appear to have exercised much influence on the theory of disease as taught in the schools, nor on the practice of the profession among the community.

SECOND LECTURE.

Remote causes of disease—Causes definite and traceable—General causes of disease—The aperient system—Treatment by increasing the quantity of healthy blood—Stimulants in debility reduce the frequency of the pulse—Alcoholic stimulants essential—Cases—Disease of animal and organic life—Cases in illustration of the two varieties of disease.

I NOW claim your especial attention to the subject which relates to the remote or primary causes of disease, than which none connected with the practice of our profession can be more critical or more interesting. If you inquire into the recent history of a person suffering from disease or illness, whatever form almost it may assume, you will learn of some event or circumstance of recent occurrence that has drawn more or less suddenly, whether on his circulating or nervous system, by which the vital powers of his body have been reduced. This condition is commonly known under the term *exhaustion*, and is the result of a sudden draught on the powers of the constitution, by which the relations hitherto existing between the physical frame and the powers of life inhabiting it, are for the time deranged.

So long as health is undisturbed by any counteracting agency beyond the natural wear and tear of time, life will be prolonged to its full term of years, but in our path through life we are liable to disturbing influences which destroy the equilibrium. These accidents draw on the vital powers and reduce them below the level of the standard health, and disease follows as an almost certain result. Let us suppose a man to consume each day a given amount of food. Accident deprives him of a portion of this food, which I will conclude equal to one half of a day's ordinary consumption. To this amount he fails in the supply of material of blood-making, and his circulation is weakened in proportion. That man has become more susceptible of disease, in a ratio with his loss of blood, than he was prior to the occurrence. Another man, under circumstances, whether of choice or compulsion, sustains an unusually great and prolonged physical effort, whether in walking, or riding, or in rowing, of which we see abundant examples in the annual folly of the university boat-race. Inasmuch as the serious consequences of these physical efforts, on which the reputation for learning of the great Universities of England appears to depend, are remote, they are not detected. I have seen several examples of disease developed in men who have undertaken a walking tour in Switzerland, and who have overtasked their powers of endurance, have returned home exhausted by the effort, and have been laid up ill for months.

Any cause which tends to reduce the quantity of blood in the system, whether by direct abstraction or by deprivation of food as the source of blood, reduces the vital powers, and disease or illness in some form is the consequence. The same principle is involved in the defective supply of oxygen to the lungs, and indeed is applicable to any and all causes that either reduce the blood in quantity, or impair its strength for circulation. Any great and unusual physical effort reduces the powers of life in proportion to its degree, more especially when coupled with loss of sleep. Exposure to cold, when severe or prolonged, is a source of subsequent derangement of health, and among other causes may be included either shock to or exhaustion of the mental powers, such as great anxiety of mind or mental emotion, and laborious and continued mental labor.

Of the above causes of disease, direct or indirect loss of blood and deficient food are the most common; and although I have above referred to them, as to also the remaining causes, as sudden, they may be more or less chronic both in their nature and in their consequences. I referred rather to sudden and severe illness or disease as the result of some sudden and palpable change in the routine of life of the individual, in which the evidence of cause and effect is immediate and conclusive. I believe, then, that in the large majority of illnesses you may trace back the malady, be it what it may, to one or other of the above incidents as its direct and obvious cause : and as certainly as one or the other of these causes has existed, you may with reason and probability look out for its result in some early attack of illness, medical or surgical, local or constitutional.

Whatever function or structure is most liable to derangement, the consequences fall upon it in the first place. It may be diarrhœa, or pneumonia, or abscess, or erysipelas, or rheumatism, or general debility. If you will adopt the habit of tracing back the recent history of any or every patient you prescribe for, laboring under positive disease, you will learn something bearing upon it of the nature I have referred to, some effort overtaxing the physical powers or reducing the circulation or nervous powers of the system; and if so, is it not a reasonable inference that *all disease begins by a stage of weakness?* The first link in this chain consists in a reduction of vital power, from which reduction the effort of the physician or surgeon consists in restoring the lost power and bringing the patient back to the standard of health.

In illustration of these principles I will relate a few cases. A healthy lady, of about 34, consulted me on account of an abscess in the breast. I concluded she was suckling a child, but I was mistaken. Anxious to ascertain the cause of so unusual an occurrence in an apparently healthy person, I made every inquiry into her past recent history. Have you had a blow on the breast? No. An attack of illness? diarrhœa? any unusual or excessive loss of blood? any great muscular effort or loss of food? All these questions she answered in the negative, and I was about to relinquish my inquiry when the lady said, " Perhaps I ought to tell you that a few weeks ago I sat up seven successive nights with a dying child." I thanked her for such important information, and the mystery was solved. Why should this lady have abscess in the breast? She had not borne a child for many years. There must be some cause for abscess, local or general. It might have been produced by a blow, but it was not. Investigation proved it to be the result of sudden reduction of her vital powers by great physical and mental effort. Why it assumed this particular form I know not. I only know that a great cause of derangement of her system had occurred. and the consequence fell on that particular organ. The late Lord R——, æt.

54, whose recent and sudden death caused so painful a sensation, became the subject of phlebitis consequent on severe bodily exercise. His pulse was constitutionally low seldom ranging above 50, and frequently not exceeding 48. There was a remarkable feature in his personal history, which consisted in a singular tolerance of the effects of alcoholic stimulants. No quantity that he ever consumed was known to have produced any apparent effect either on his spirits or on his pulse. The daily consumption of two or even three bottles of wine left his pulse as it found it, at 50 or 52. In the two months during which he was under my care, I prescribed stimulants to the full extent of prudence; but he was as calm and as free from all indications of excess in the evening as in the morning. His appetite for food was large, and he indulged it. If I reduced his wine or his brandy by the substitution of medicated stimulants—ammonia, ether, etc.—he flagged. By the frequent resort to wine in reduced quantities I raised his pulse to 56. Throughout his attack he often assured me that he had never had the sensation of better health in his life, and he only desired my permission to get out and take exercise. His left leg was first affected, which passed through all the stages of obstructed venous circulation in about five weeks: consolidation, pain over the track of the larger veins, œdema of the entire limb, and, finally, softening of the veins, and absorption of the fluid. The right leg then became the subject of the same disease, and, under the same treatment, it progressed satisfactorily to its final stage. His confinement to his couch was limited to the additional term of ten days, when I assured him he should take a drive. On this information, and in the delight he felt at the termination of his long and severe confinement, he stuck his thigh a smart blow with his open hand, to prove to me how free his limb was from pain. This occurred late in the afternoon. At half-past ten he had some difficulty of breathing, and he broke out in a profuse perspiration. Stimulants availed him not, and at half-past eleven he was dead.

Here is physical exhaustion as the cause, acting on a system characterized by an unusually low circulation—phlebitis the effect, and sudden death by *embolism*.

Inflammation of the uterus, or of the venous system below it, follows hæmorrhage in parturition. Accoucheurs with large experience will confirm this fact.

Many years ago I was present at the attempt to reduce a dislocation of the femur in a London hospital. The man, who was to all appearance in perfect health, was bled largely from both arms and from the temporal artery. He also took tartar emetic to the extent of some fifteen grains or more. The quantity of blood abstracted from his circulation must have rendered the functions of the heart a sinecure, for it amounted to 120 ounces. He survived a week, and died of phlebitis, as I had predicted.

I might multiply these examples, but it is unnecessary if you comprehend the force of the few cases I have quoted.

The causes of disease I refer to above may be classed as follows:

1. Loss of blood, whether accidental or effected *at the hands of science*, catamenial, hæmorrhoidal, etc.

2. Loss of food, or failure of the material of blood-making.

3. Excessive purgation, whether natural and spontaneous,

as in diarrhœa, or artificial *at the hands of science*, carrying off nourishment.

4. Breathing impure air, by which the quality of the blood is deteriorated.

5. Loss of sleep.

6. Excessive muscular effort, as in walking great distances, boat-racing, etc., as in the annual University struggle, of which it may be said the higher the rank of the antagonists the greater the danger.

7. Extremes of temperature, especially of cold.

8. Great mental emotion or mental shock.

9. Protracted anxiety of mind.

Of the above sources of subsequent illness or disease, the influence of the first four is received more directly by the vascular system, the five latter either partially or entirely by the nervous.

Before speaking of the general treatment of disease, to which I shall shortly come, I wish to say a few more words on the subject of some prevailing doctrines which are so generally adopted by our profession, and the entire soundness of which appears to me yet open to question and inquiry. Why do you order an aperient in nearly every case of disease you are called to see? If the patient is strong and vigorous, the dose is a full one; if weakly, a milder form; but always an 'aperient; and generally it is combined with mercury, which indeed forms the staple of the medicine all but universally prescribed. As we commonly go at once to the cause in our treatment, and prescribe our best remedy, the natural inference is that there is either liver or intestinal derangement as the cause of the disease. You say you desire to unload the liver and to remove extraneous and irritating matters from the intestinal canal; and very good treatment too, if the liver is at fault and the alimentary canal demands that kind of relief. But what evidence have you of it? You say the liver is congested. It is a term in the mouth of nine tenths of the profession practising medicine and surgery in the dominions of her Majesty the Queen. The practitioners on the Continent of Europe take a different view of these matters, and not without some show of reason, for English doctors, though I entertain the highest respect for their attainments, do not monopolize all the knowledge of the world. In discussing the subject, I am prepared to acknowledge the occasional presence of constipation as the result of torpid action of the bowels, in which condition the liver may or may not be involved; but I am myself unable to detect what appears so obvious to many others—a congested state of this organ, calling for large doses of chloride of mercury, supposing that form of drug to be the best corrective of the evil.

As I believe a person who is the subject of disease, itself so commonly the product of exhaustion, should not undergo further reduction of his strength without a good and sufficient reason, and as I doubt the congested state of the liver and find the torpid condition of the large intestine, if it exist at all, an evil on a small scale, I prefer to look to the disease itself, (suppose it, if you please, erysipelas or any other malady,) and if the pulse is soft and compressible, indicating distinct constitutional weakness, *and not otherwise*, I prescribe at once a tonic remedy. I look upon this torpid condition of the large intestine—for, observe, nearly all constipation is limited to this part of the alimentary canal—as merely a symptom of the general debility, and is increased by loss of appetite and the absence of food, and is not to be rudely and violently assaulted by drastic purgatives; and one of the first signs indicating the sound principle of a tonic treatment will appear in the gradual, but certain, restoration of the functions of the alimentary canal to a state of health. Constipation is very commonly, though not invariably, the concomitant of weak health and low vital power, and is caused by defective power of the muscular fibres of the large intestine, which are, when compared to the small intestine, very limited in quantity in relation to the size of the intestine. You treat constipation by means of purgatives which act on the mucous membrane only, I prescribe iron to give tone to the muscular coat. Your treatment affords a temporary and transient benefit; mine is a permanent one. Which will you prefer?

And this subject leads me to add a few words on the treatment of disease adopted by my own respected teacher, Mr. Abernethy, who upheld the doctrine that health demanded the daily action of the bowels, and that the liver was an offending organ requiring the daily administration of a stimulant or a supposed provocative to the secretion of an increased quantity of bile. No doubt Mr. Abernethy's treatment was to an extent successful, because, as the tendency of the large majority of diseases is toward recovery at the hands of Nature, his cases prospered like those treated on other principles. In these days one does not quite comprehend how a man of Mr. Abernethy's grasp of intellect could delude himself with the idea either that disease could so generally depend on torpidity of the liver, or on imperfect action of the muscular coat of the large intestine. If it be necessary to obtain as the requisite condition of health a daily action of the bowels, why did Nature give us large intestine sufficient to contain *at least a week's consumption of food!*

In cases of extreme weakness good service is often rendered by "locking up" the bowels, and reducing their action to that of alternate days.

I consider the treatment of the great majority of diseases to consist in increasing the quantity of healthy blood and giving force to the action of the heart. *You can't cure disease with a feeble pulse.* Mend the pulse, and Nature will do the rest of the work. On this principle disease in general may be treated, so far as my observation has gone, with pre-eminent success. In order to appreciate fully its force, you must start with the conviction that Nature cures and not man : man removes obstructions from her path, and nothing more. This done, he awaits the onward move of the great machine, like to a great ship of gigantic weight, which, quietly held in her position at rest by a few timbers, immediately obeys the great natural law of gravitation on their removal, and glides into the water below. Did man launch this vessel, or did Nature? With as much title may the physician or the surgeon declare that he cured a disease. There are of course occasional exceptions to this assertion in some cases of operative surgery.

The object of treatment is to restore the pulse to its normal standard of force and frequency. Give it due force, and the heart will determine the number. As a rule, in cases of debility, it is too frequent, and frequent, because the quantity of blood in the system is below the standard of health. Increase the quantity and the pulse falls. Assure yourselves of this. Unthinking persons jump to the conclusion that brandy or other stimulants necessarily raise the pulse, but this supposes that we start with a healthy pulse at par. I am talking not of health, but of disease. In my capacity of Examiner at the College of Surgeons, I often put this question: "if you take a pint of blood from a healthy man of 40, with a standard pulse of 68, what effect will be produced on the number of pulsations by the loss?" What do you imagine is the frequent reply? "It reduces the number to 60!" And this curious answer explains something of the phenomena of venesection so universally practised some years ago, when in reporting on a case it is said, "His pulse rose on bleeding, and so I bled him again." As a rule, you will find that, whenever the frequency of the pulse is above the standard of health, *as an indication of debility*, a stimulant will reduce it. I tried this experiment, or rather I obtained this test, for it was not an experiment On coming out of a Turkish bath of something more than the usual intensity of heat, my pulse had risen to 90; I drank about two ounces of wine, and my pulse fell to 75 within a few minutes.

I have something more to say on common matters relating to the routine of every day's practice. There is great repugnance prevailing in the medical mind to the employment of stimulants. It is, perhaps, mainly founded on the moral evils of excess. Inebriety is a low, vulgar vice, and therefore the agents producing it are morally objectionable and medically injurious; but you must draw a distinct line between health and disease. In health I am the advocate, both in precept and example, of moderation, in disease not. I think one reason why stimulants are so little resorted to in cases of debility is because, being administered in hesitating quantities of doubtful utility, their value is not appreciated. You must not gauge the capacity for alcoholic stimulants in disease by the capacity of the same person in health. Those only who adopt this treatment as a principle are cognizant of the remarkable tolerance of stimulants under great prostration of the vital powers, and when required as the antidote to prostration I maintain they are perfectly harmless. It is the remarkable tolerance of alcoholic stimulants in these persons to which I wish to direct your earnest attention. You can't intoxicate these people; you cannot even unduly excite them. It is a common remark among observant persons under treatment, "I drank three glasses of wine, but it had no more effect on me than so much water;" whereas less than the same quantity during health would have produced partial intoxication, or at all events excitement. These are the persons who demand the free and fearless resort to stimulants, and to whom alcohol is life.

Do not, I again urge upon you, measure the quantity to be administered by the glass. Gauge it by its effects, and so long as weakness prevails, indicated by the pulse, persist in its use; and as you proceed in your treatment, and the services of the wine become more and more palpable, you will not fail to see that its consumption becomes less and less essential to onward progress, and in the course, it may be of months, of weeks, or even of days, you will reduce the quantity to the standard of health, for the capacity for wine has passed away with the disease that claimed it. You need never listen to the objection, often urged, that the resort to stimulants will become habitual; be assured it is entirely groundless.

Now listen to the following examples of this principle of treatment. A young lady whose daily consumption of wine has never exceeded two glasses, oppressed by the heat of a crowded room, faints. The heart and the brain have failed in their functions by reason of the imperfect oxygenation of the blood. Recovering from the first stage of sudden depression of her vital powers, her symptoms pass into those of hysteria. We have all witnessed cases of the kind. Ordinary and mild stimulants are unavailing, and make no impression. An hour, two hours, have elapsed, and she is yet prostrate. Before her recovery this girl has consumed three wine-glasses full, or upward of half a pint, of brandy and other cordials, and often more. What is the explanation? What is the physiology of this? Had you administered this quantity to your patient during health you would have done her a serious injury. She now takes it with benefit and with absolute impunity. The shock to her system was for the time dangerous; the stimulant, given at another time, would have proved dangerous also; the two united are harmless. The one has balanced or antagonized the influence of the other. There remains neither headache, nor nausea, nor any evidence of excess, because all the stimulating power of the brandy was required and consumed in re-establishing the circulation.

Again: a gentleman of abstemious habits, ill clothed for the occasion, ventured in a cold season of the year, to mount to the summit of a mountain in Wales. He was seized with intense and sudden prostration from the effects of the temperature. His suffering was severe. At this opportune moment, a friend supplied him with a half-pint bottle of raw brandy, every drop of which he drank at a draught, as he would have drunk a glass of table beer. The only effect produced on his system was warmth. He drank more brandy at that moment than he had ever before consumed in a week. No evil resulted. The cold and the brandy being in antagonism, each entirely neutralized the injurious effects of the other. I need not say that a similar draught, unprotected by the depressing influence of the intense cold, would have proved almost dangerous to life itself—certainly most injurious to health.

I attended in his last illness the late Duke of ——, a man of remarkably abstemious habits. His malady was a mortal one. Some months prior to his death he became the subject of frequent attacks of rigors, which continued with great severity for two or three hours, and left him in a condition of extreme prostration. I called one day opportunely, and found him laboring under one of these attacks. I gave him two thirds of a wine glass of raw brandy without effect. In five minutes I repeated it; and he took a wine glass full of brandy every seven or eight minutes till the attack passed off. He had then consumed three quarters of a bottle of pure brandy within an hour. I saw him in four hours afterwards, apparently well. In reply to my question, " How do you feel ? " he answered,

"Well, I think you gave me pretty large doses of brandy, but I cannot say I feel in any respect the worse for it." Nor was he. Throughout his subsequent illness he had no attack of rigors. The prostration of the vital powers in this case must have been immense.

One of the last cases attended by a late eminent physician—now, unhappily for the cause of Medical Science, lost to the profession—was a gentleman of about 30, who was the subject of hydrothorax. His case was urgent, and I was requested to see him. Concurring in opinion with the physician in attendance on the necessity of giving immediate relief by operation, I punctured the chest and got away six pints of serous fluid. The cavity thus made was entirely occupied by atmospheric air, to the admission of which I have never yet ascertained the grounds of objection, or seen any injurious result therefrom. I have done this operation often, and have never considered it desirable to exclude the air from the cavity within. Thought irrelevant to my subject, I cannot forbear asking the simple question, What is supposed to occupy the space from which the fluid has been removed? Certainly the lung will not immediately expand, and the walls will not collapse for a period of weeks or months. At the conclusion of the operation the patient was much exhausted, with a pulse of 130. It was a great object in the treatment to reduce it. What means would you have adopted? What does professional usage demand? Salines, think you, or purgatives? No, on the contrary. He took the only remedy that could have probably saved his life. He was ordered one ounce of brandy diluted with an equal quantity of water every three hours, and on the day following his pulse had fallen to 90, and he recovered. If this treatment was not sound, it ought to have proved fatal.

I have abundant examples at hand of the efficacy of this principle, but I have quoted sufficient for my present purpose. They go to exemplify this therapeutical principle, namely, 1. That stimulants alone can restore the vital powers under great and sudden prostration; and 2. That under great and sudden prostration the capacity of the system for stimulants is enormous, and that they may be administered to almost any amount with safety.

Now, gentlemen, in the prosecution of your profession, while engaged in the study of by far the most critical department of medical education—the nature of disease, or *diagnosis*—have you ever thought of the important distinction which exists between the diseases of *Animal* and *Organic* life? And yet they are worth considering. A few years ago I made some calculations on the subject of the relative quantity of blood supplied to each. I have not these calculations at hand, but the quantity of blood so distributed is nearly equal, though rather the larger proportion supplies the structures of *Animal* life.

The nerves, as you know, are essentially different. *Animal* life being spinal, *Organic* life, ganglionic. If we take the whole material of the body into the calculation, the structures of *Organic* life do not exceed one tenth of the actual weight, yet they receive nearly one half of the whole quantity of arterial blood.

The "two lives," as they were termed by the great French physiologist, Bichat, are more or less independent of each other; both in health and in disease. The diseases of *Organic* life are more serious, because they involve structures more directly essential to life itself, and the derangement of any one of them involves the whole system more thoroughly, than does a similar amount of disease in the structures of *Animal* life.

You will hear a man say "he can't have much wrong with him," because he is capable of great exercise, and can walk any reasonable distance. This is a bad test. So long as the muscular system is not directly involved, and sufficient blood is made to supply both it and the spinal cord, from which it derives its nerves, a man may labor under a considerable amount of disease of the structures of *organic* life without his muscular system being involved. When we speak of a man's "health" in general terms, we refer instinctively to the functions of the interior of the body or of the organs, not of the muscular system; we refer to the functions of the brain, to respiration, to circulation, to digestion, and the assimilation of food, etc. The muscular is but subservient to these other organs which really constitute life itself.

You will find this subject, which makes palpable the broad distinction between the two classes of diseases, one of the most interesting connected with the study of your profession. It marks the subsidiary character of the muscular system, and of the structures subservient to it, as holding secondary rank in the general economy of the body, while its deviation from health influences the economy in a secondary degree only. You can form no true gauge of a man's capacity for muscular effort, however perfect may be all his attributes of health, any more than you can rightly infer the soundness of his organic system because his capabilities of prolonged physical effort are great. Take the case I have above quoted of Lord R——, a man capable of unusually great muscular exertion, and who could walk for hours without fatigue, while his heart and arterial system betrayed a condition of weakness which suddenly terminated his life.

A yet more remarkable example is found in the case of a young gentleman who died some few years since, whose muscular system was the subject of a form of paralysis, first described by Dr. Meryon in a paper in the Medical and Chirurgical Transactions, and subsequently by Cruveilheir, in which the muscular filaments of animal life failed in their development throughout the body, while all the organs within the body, and the organic muscular fibres subservient to them, gave evidence of perfect health. I have a young gentleman under my care at the present time whose physical frame is characterized by remarkable muscular power. His strength is prodigious. But he has a feeble pulse, is very susceptible of cold, and is subject to fainting fits that are all but alarming.

But it is scarcely necessary to refer to individual examples which proclaim the subservient nature of the muscular to the *organic* system of our frame. It may be seen at every step of our professional lives.

THIRD LECTURE.

ON HYSTERIA OR GENERAL AND LOCAL NERVOUS IRRITATION.

Imitations of real disease—The study of, much neglected by the profession—Indicated by local pain and muscular spasm—Distinct from inflammatory disease—The term " Hysteria" objectionable—Nature and source of Hysteria—Errors in diagnosis—Critical examination of cases—Hysteric affection of joints---Hysteric affections of the spine—Hysteric contraction of fingers.

In the whole range of practical surgery there is, perhaps, no one subject that claims your earnest study more important than that which I have selected for this and the following lectures. It is not a question of diagnosis between two diseases more or less resembling each other. It is a question of disease or no disease, of reality or imitation, of true or false; of whether your purgatives, your bleedings, sweatings, irritants and counter-irritants, and your whole battery of antiphlogistics, shall be launched against a true disease in the flesh or its ghost; whether you are to contend with a reality or a shadow. This absence of discrimination between two conditions of disease and no disease are painfully frequent among medical men, especially among those to whose charge is assigned the care of local and surgical diseases. "In one shape or another," observes the greatest of modern surgeons, "you will meet with them at every turn of your future practice."

It may be asserted with truth, that every part of the human body supplied with nerves, be they cerebral, spinal, or ganglionic, may become under provocation the seat of local symptoms so closely resembling those of the real disease to which that part of body is liable, as to appear identical with it, and the resemblance to which is so perfect as to deceive the best of us. They are not cases of occasional or rare occurrence. They come before us in the daily and hourly walks of professional life. They monopolize a share and not a small one of all cases under treatment, whether medical or surgical, but the latter predominate. The closer you scrutinize them, the more penetrating your inquiry—looking into and not at them—the more perfect will be your diagnosis, and the more will you be astonished that a form of disease so remarkable and so common should have hitherto occupied so little of your thoughts.

It is well to call your attention to this description of malady at the early stage of your professional career. Many men pass through life, engaged in active warfare against disease, on whose convictions this variety has scarcely dawned. And this is a truly remarkable fact, which owes its existence to the predominating influence which the heart and the arterial system exercise over the judgment of the profession at the expense of a system yet higher in the scale of organization, more sensitive, and far more liable to morbid impressions, namely, the cerebro-spinal nervous system.

Whenever a new case of disease presents itself to us, we jump to the old doctrines of inflammation, we talk of congestion, and of capillary action, and of deposits of lymph, and we refer the attendant pain and heat to an inflammatory condition, of which the local nervous derangement is an ordinary symptom. We should endeavor to assign to each system its proper place in the pathological scale, and to discriminate more accurately than is generally done, the indications which belong to the morbid conditions of each, whether existing in combination or separately. For be assured they do exist, both separately and in combination with each other. You may have varieties of inflammation in which the local pain is trivial when compared with its severity in other cases; while, on the other hand, examples daily occur in which local as well as general derangement of the nerves, whether of the part or of the whole body, exists as a morbid condition entirely independent of the vascular system. Nor is this derangement confined to the sensory nerves. If we have local pain as the indication of excessive activity of the nerves of sensation, we have spasms and convulsions indicating derangement of the nerves of motion, each of which, or both may prevail without heat, or redness, or swelling. We may daily see severe forms of nervous exacerbation without the slightest corresponding increase of action of the vascular system. There is this important difference between the morbid states of the vascular and nervous systems, that while local inflammations are dependent on local causes, aggravated only by the impaired condition of the general health, local nervous diseases for the most part originate in the centres of nervous power, the effects of which are exhibited in remote parts of the body—it may be in a pain localized in a given spot, whether on the surface or in deeply seated parts, which to our senses holds no especial relation to its nervous centre; it may be in a temporary, or spasmodic, or permanent contraction of the voluntary muscles bending the joints of the extremities in permanent flexion, or obliquely drawing the head upon the trunk, or involving the whole motor system, as in tetanus. No known nerve that conveys sensibility from its centre to its periphery, no motor nerve that carries volition from the brain or spinal cord to a voluntary muscle, is exempt from this morbid tendency.

The vascular system, consisting of arteries, capillaries, and veins, has its own special diseases peculiar to the structures engaged in the circulation of the blood. The attendant symptoms are heat, redness, pain, and swelling, the latter symptom being due to a separation from the capillary system of some constituents of the blood, whether in a fluid or solid form, while the morbid conditions of the nerves and the structures in which they originate are characterized by simple aggravation or excess of the functions of the nerves affected, the natural sensibility of the sensory nerves running into pain, and the moving power of motor nerves into convulsions, or spasm, or permanent contraction. In diseases of the vascular system we have changes of structure, in the latter not. It is necessary to make very clear the line which separates the two classes of disease, lest we fall into the common error of applying to both the remedial agents which are applicable to one only.

The diseases originating in or involving the vascular system we treat locally by various agents, leeches, blisters, etc.; in diseases confined to the nervous system these local remedies are useless and even injurious, and we treat them through the constitution. In cases of *Tic* do we derive benefit from leeches, or blisters, or from other form of depletive agents? Assuredly not.

Now the disease which forms the subject we have to consider belongs to the nervous and not to the vascular class; and I select from this variety that occasionally known under the term "Hysteria," than which no name can be more inappropriate or objectionable. It may well be doubted whether, except under very occasional circumstances, such a relation holds between the *uterus* and this remarkable train of symptoms as to justify the employment of the term Hysteria. In the large majority of cases there is no connection between them beyond that which the disease holds with the other organs of the body. In the name of a disease we are supposed to recognize its form and nature, whereas the term I have quoted conveys to the mind no distinct idea of either one or the other. And there is a positive objection to the resort to it, in the fact that the word carries with it the association of a malady of small and insignificant dimensions, while the malady itself is of great magnitude. We associate with it the idea of "hysterics" and "vapors," as they were formerly called. I wish to raise your attention to the level of a great malady, and not of a trivial derangement of the hour. I remember a law case in which the counsel challenged a medical witness as to the name of the disease, and he replied, Hysteria. "Hysteria!" said the learned counsel, addressing the jury; "we all know what Hysteria means. My client has come into court to obtain compensation from a jury of his country for a permanent injury, by which all his prospects are blighted, etc., etc., and the gentleman in the witness-box, with no sympathy for his misfortune, proclaims the disease to be that of trumpery hysterics;" and the jury, with rod in hand, let it fall heavily on the defendant's back.

But there is a more solid objection than these, namely, that it is founded on a false pathology, by the employment of a term that conveys an impression of its source and nature founded in error. The disease consists in the local evidence of some irritation or derangement of one or the other of the nervous centres of the body, namely, the brain or the spinal cord; at least, such is the received pathology. But the subject is a very obscure one. We have no very definite idea of what we mean by "irritation." We all employ it, and so general is its use that I do not know how we can get on without it. "Irritation of the nervous centres" is a useful and not an ill-sounding phrase, though somewhat mysterious, but it is no reflection on medical science that we cannot explain all the phenomena of life; and as the term is somewhat wide in its application, and does not commit its employer to any very defined opinion on obscure matters, on which it is very difficult to form any opinion at all, I presume we shall retain it.

One good reason that may be assigned for the persistent employment of the term Hysteria—a term we all know to be objectionable—is the difficulty of finding a substitute for it. We call the disease "local nervous irritation." It is "exalted nervous sensibility;" but in naming a disease so definite as this we require a term equally pointed and definite with the thing itself. That we have not got.

Sir B. Brodie says, " I employ the term Hysteria because it is in common use, but the etymology is calculated to lead to great misapprehension."

Failing the name, let us look to the thing, and if it be so critical as I have assured you, let us attach to it the grave importance its frequency and its magnitude demand.

It may be asserted with truth, that *every part of the body may become*, under provocation, *the seat of an apparent disease that in reality does not exist;* that it may and often does assume all the attributes of reality with an exactness of imitation which nothing short of careful and accurate diagnosis can distinguish from the real disease. You think this impossible. Surely you know a diseased knee-joint, you reply, when you see it. You find severe pain, aggravated by the slightest movement. The temperature of the joint may be raised, and it is slightly swelled. You leech, you blister, you employ an iodine liniment (few cases escape it), you may even resort to issues, but the evil remains in spite of all your remedies, which have been applied to the wrong "system." It is the nervous, not the vascular, that is involved, but the nervous has imitated the vascular and deluded you, and led to the employment of false remedies, which have failed to reduce the pain or give mobility to the joint, and the general influence of which on the health of the patient cannot be said to have proved eminently serviceable.

The case on more perfect investigation, proves to be one of local nervous irritation, or Hysteria. You think you will not be again deceived, but you are mistaken. A single error, corrected by the experience of another, will not teach you Hysteria. You are consulted by a lady in reference to a daughter eighteen or twenty years of age, who has exhibited failing health for some time, and now complains of her inability to walk in consequence of a pain in her back. You examine her, and discover that she suffers extremely on pressure over two or three of the lower dorsal vertebræ, or on any other of the twenty-four. You repeat the examination with the same result, and you make a report to the mother that her daughter has "*spinal disease.*" The result of your opinion is two or more years' confinement to her couch, coupled with the usual concomitants of restricted diet, alterative and other depletive medicines, leeches, blisters, and issues. Suppose those structures which you have declared to be the seat of organic disease to be examined under a microscope, what would you discover? Nothing. There is no disease whatever. As the nature of this malady dawns upon you, now awakening to a conviction of its frequency, you resolve to be more wary in your future diagnosis.

You are now consulted by another young female patient on account of a tendency in one or more fingers to close in flexion. In the attempt to straighten them you cause intense pain, and if persisted in the consequences may be serious. Your patient appears in fair average health, and all her functions are regular and healthy; while the hand, for all ordinary purposes, is useless. Under the idea that she may have some chronic inflammation of the theca or of the palmar fascia, you treat it with the usual remedies. But your remedies produce no impression on the finger, which continues obstinately flexed as before; you adopt another principle of treatment founded upon a more correct diagnosis, and your patient recovers. These cases sound strange to your limited experience. You think they are rare, and brought forward from a distance, and with an effort. By no means. *They are cases of daily occurrence.* If you could suddenly throw off that nebulous vision of vascular disease which years of bad pathology have impressed upon your judgment, you would see them in their true light. You may deem them to be exceptional. I assure you they constitute the rule of disease, and not the exception. *Real disease is the exception.* Speaking of one variety, and they have all characters in common, Sir B. Brodie, a man who rarely committed an error in diagnosis, says: "I do not hesitate to declare that, among the higher classes of society, at least four fifths of the female patients who are commonly supposed to labor under diseases of the joints, labor under Hysteria and nothing else." I would venture to enlarge this statement as regards the "upper classes," by including a large proportion of the lower; for much of my own experience of Hysteria has been obtained from the wards of St. Bartholomew's Hospital, and in reference to spinal affections in young persons, I unhesitatingly assert that real disease is not found in a greater proportion than one case in twenty, and even this is a liberal allotment.

Have you never experienced the difficulty of discovering an object floating on the air, such as a bird singing overhead, or an early star in the evening? When once the object becomes visible, the eye is readily adjusted to it, and when you look again in the right direction, it is the first object that strikes the eye.

And so with this class of diseases. They are not seen, because they are not looked for. If you will so focus your mental vision and endeavor to distinguish the minute texture of your cases, and, as I have said, look into and not at them, you will acknowledge the truth of the description, and you will adopt a sound principle of treatment that meets disease face to face with a direct instead of an oblique force, which far too generally claims the credit of a success for which nature alone is responsible.

I have selected above three varieties of this local hysteric affection. Let us consider them a little more in detail, with a view to detect the fallacy which classes them under diseases of the first or vascular division, by which I mean an abnormal condition of the blood vessels, leading to changes of structure or, altered relations of the parts, whether by suppuration, or ulceration, or fibrinous deposit, or local death of the tissues involved. In the first case the knee is the seat of pain. The subject is a young female. What evidence do we commonly look for when the joint is really diseased? We look first for a cause. Diseased joints do not occur without a palpable one, and particularly in young persons. There has been no violence, no fall or blow, to which to attribute it. Had there been, the nature of the disease is obvious enough. There is no considerable increase of heat, and if inflammation is present perceptible increase of heat is constant. There is no effusion into the joint; the form of the articulation is unchanged. The pain and the immobility or stiffness of the joint remain, notwithstanding your remedies. Local depletion relieves the pain of inflammation, but not of Hysteria.

But you persist in your principle, and the depletive treatment is continued, and thus months elapse, yes, even years. I was once told by a young lady that she had applied twenty-seven blisters to her knee-joint, from which she could not say she had derived any benefit. Now it ought to be obvious that, if a painful joint occurring in a young female without local cause is unaltered in form or size, and is free from heat or redness, and that the chief and almost the only symptom, that of pain, varies in degree at different times, and is fluctuating in character, the disease is not of the inflammatory class; and if not it must be nervous, and you can't cure pain with leeches. You know that pain alone, which consists in an exalted nervous sensibility, does not constitute what we strictly understand by the term *disease*, although we apply it generally to any deviation from health, whether local or constitutional. At length the truth is brought home to you. You change your treatment by the substitution of local sedatives and general tonics, and your patient at once moves forward in the direction of recovery.

Take the second case. You have declared your opinion that this girl is the subject of disease of the spine upon the single evidence of local pain produced by pressure of the fingers on the spinous processes of the vertebræ. It has escaped your observation that this pain is equally severe whether pressure is slight or not. In fact, the degree of pain indicated by either writhing or exclamation holds no relation to the force of the pressure made. The slightest touch creates as much suffering as the greatest pressure of the hand, and *often more*. It is on this evidence alone you have founded your opinion of disease of the bony structure of the spinal column. It is on this evidence you have consigned this young lady to two years' confinement to her couch, to the loss of education, to restricted social and domestic intercourse with her family and friends, and to much moral and physical suffering. Now when you talk of disease of the spine, what do you mean? What *structure* is diseased, and what form of disease is present? Is it seated in the *body* or in the *processes* of the vertebræ, or in the entire bone? and what description of disease has invaded the particular vertebra of the twenty-four? Is it inflammation, or caries, or necrosis? Caries you will say; and you select this form because, and only because, you know the spinal column is the subject of carious disease under conditions favoring it. But there is this remarkable feature in carious disease of bone well worthy of notice, namely, that it is almost destitute of pain, that there exists no relation between the extent of the disease, which may be great, and the pain attendant on it. It is not like *inflammation* of bone, whether simple or severe; nor does it resemble *necrosis* or simple death of bone. Presuming this statement true, can you in reason feel satisfied with the evidence of disease obtained by manual pressure? Then, again, where is the disease situated? what is its precise locality? If in the body of the vertebræ, is it not almost absurd to suppose you can detect it by the slight pressure of the finger *on the summit of the spinous processes*, which are themselves rarely involved?

Fifty or sixty years ago a provincial surgeon of some note recommended the application of a hot sponge to the spine, with a view to detect disease of the bodies of the vertebræ. There was some excuse for ignorance on this subject at that time; there is none now. Of all the fallacies that cling to professional practice, of all the false doctrines which the pardonable ignorance of a former generation has entailed on modern surgery, none can surpass that which affects to detect carious disease of the body of a vertebra by drawing the fingers down the spine. It is only not ludicrous because the consequences are so serious to the victim. It would be a bold assertion that such morbid changes in the spinal column cannot occur; but I do think humanity would be a gainer if all teachers concurred in asserting that they *could not*, so rare is the real disease, and so palpable to the eye when present. Suppose a young person in moderately good health, and occupied in daily exercise, complained of a pain in the condyle of the femur, without any other indication, should you be warranted in declaring she had serious disease of the bone? Look to the functions of this important vertebral column; how is it possible it can support the body in the upright posture if one or more of the component bones of the pillar are destroyed? And yet I have known many examples in which the subject of this imaginary disease has joined a party and danced for the whole evening. One wonders that such a person did not drop into pieces! For myself I candidly declare that I have scarcely ever seen a case of true disease of this form. I can bear testimony to spinal affections and destruction of bone to any amount in psoas or lumbar abscess, or in angular curvature, or to damage done to the column by local injury; but to these suppositious cases which exist only in the brain of the surgeon, I am a stranger, and if they exist otherwise than as rare examples of spinal disease I have much to learn. Have you ever seen a person recover from actual disease of the spine? I do not mean to infer that death inevitably follows, though that result is by no means uncommon; but I allude to recovery without some distortion or some permanent evidence of past disease. And yet you may be surprised when I assure you that all these young people recover sooner or later: *sooner* if the surgeon in attendance is familar with hysteric affections, *later* if he is not. Thirty or forty years since, these cases were, happily for our time, far more common than at present. At that date, and for how many years anterior I know not, all the sea-side towns were crowded with young ladies between seventeen and twenty-five years of age and beyond it, who were confined to the horizontal posture, and were wheeled about on the shore in Bath chairs, on the supposition that they were the subjects of spinal disease. They were placed under much medical and dietetic discipline, not of the most invigorating character, and the large majority carried a pair of handsome issues in the back. Brighton, Worthing, Hastings, and other places on the South Coast were largely tenanted by these unfortunate females, to which a moderate sprinkling of young gentlemen was added. What has become of all these cases? They appear to have vanished just in proportion as the eyes of the surgeon have opened to the absurdity of inferring that pain alone, which locates itself with remarkable precision in Hysteria on a given vertebra, can indicate the presence of organic disease of the body of the bone without collateral evidence in its favor. When the spinal column is really diseased the case is obvious at a glance; the health is degenerated, and the whole system proclaims to the eye of the surgeon the presence of a great evil. These examples are but a miserable mockery of the reality, and a fraud on the judgment of the ignorant.

With regard to the third example, that of permanent flexion of the fingers, it is apparently so truly local an affection that there is some excuse for error, but only because hysteric affections are not half studied. When one or more of the fingers is permanently flexed from local causes, the seat of disease will be found in the fascial structures of the hand or in the finger itself, or a joint may have been diseased or dislocated; but here there is no thickening, nor hardness, nor other morbid change of structure. The finger is simply bent, and the attempt to straighten it is painful. The cause of this morbid condition of the flexor muscle is referred to its nervous centre placed in the cervical portion of the spinal cord.

There are some curious phenomena connected with this form of Hysteria to which I shall call your attention in the next Lecture.

FOURTH LECTURE.

Symptoms of common paroxysmal Hysteria—Constitutional liability—Nerve and nervous system—Effect of railway accidents—Influence of the mind—Effects of an unstrung nervous system on the actions of daily life—Imitative or contagious Hysteria—Surgical cases of—Distinct from Neuralgia—Relation between Hysteria and the brain or spinal cord—Hysteria combined with real disease.

ON HYSTERIA, OR GENERAL AND LOCAL NERVOUS IRRITATION.

IN the last Lecture I referred especially to Hysteria in its local forms. You need scarcely be told that it is a disease involving the entire constitution. It is not my intention to enter at length into the detail of this more common form even were I capable, but I have something to say on this head that may be useful to you. You will not believe that either the mild or the severe and paroxysmal attacks of hysteria occur without a cause. That cause may be traced to constitutional weakness or previous exhaustion in some form or other, fatigue of body or mind, or great mental emotion, causing a shock. Then follows at different intervals of time the attack, attended by headache, nausea, pain in the back—*globus hystericus*, as it is termed—uncontrolled muscular contraction, convulsion of the diaphragm, indicating by fits of crying, sobbing, laughing, its close relation to mind, violent palpitation of the heart, fixed pain under the short ribs, increased secretion of urine; not uncommon in all nervous affections, especially of women. Such are the predominant signs of general hysteria. It is a medical disease, not a surgical one, and I dare say you are as familiar with the general symptoms as I am. The great Dr. Sydenham has testified to the general prevalence of this disease in all classes of society. He says, "The frequency of hysteria is no less remarkable than the multiformity of the shapes it puts on. Few maladies are not imitated by it; whatever part of the body it attacks, it assumes the appearance of the disease to which that part is liable." I don't see how this striking truth can be told in stronger language; and again I tell you the mock is far more common than the real disease, and I warn you against that error in diagnosis which confounds the one with the other.

It is not an easy task to select the class of constitutions most liable to hysteric disease. Probably under certain conditions of impared health the large proportion of the community would give evidence of its presence. Certainly it is uncommon in lower classes of males, and among those who occupy the beds in our public hospitals. We know, on the other hand, that it is most prevalent in the young female members of the higher and middle classes, of such as live a life of ease and luxury, those who have limited responsibilities in life, of no compelled occupation, and who have both time and inclination to indulge in the world's pleasures—persons easily excited to mental emotion, of sensitive feeling, often delicate and refined. Such are among the mental attributes of hysteria. But hysteric diseases are not confined to the young. I have seen many examples in females of 40 and 50. Do not imagine hysteria to be a disease peculiar to persons of weak minds. It will often select for its victim a female member of a family exhibiting more than usual force and decision of character, of strong resolution, fearless of danger, bold riders, having plenty of what is termed *nerve*. If you tell such young people they are nervous, they take offence, because they misinterpret the meaning of the word, and so may you. And they may well misinterpret it, for, like the word "irritation," its popular meaning is both various and indefinite. It is essential that we attach a definite idea to this term in its different applications, and I must digress for a moment to endeavor to explain them.

I have already used it one sense.

1st. The word *nerve* is employed to express the mental condition of vigor, boldness, and resolution, as when a man's nerves strung up to meet danger. So ladies are said to ride with more or less "nerve."

2d. We employ it in its physical sense as a part of the general system of the nerves of the body.

3d. We have the term "nervous system," which may be not inappropriately defined as holding the same relation to the "system of nerves" in its physical sense, as the physiology of a part holds to its anatomy. Yet this definition is imperfect, because the properties of the nerves of the cerebro-spinal system, with the small exception of those of specific sensibility, such as sight, smell, taste, etc., begin and end in the functions of motion and sensation. By the term "nervous system" we understand the general influence which the nerves in a physical sense exert on the constitution, the healthy or tonic condition of which is in a ratio with the combined force of the two systems, namely, the arterial, or circulation of blood, and the nervous. Health depends on the coincidence of these systems in perfect action. If the circulation fails in power, the consequence of this weakness falls on the nervous system, which is dependent on the circulation for its health and vigor. Failing an adequate supply of blood, this system is unstrung, and morbid sensations, endless in variety, take the place of real; and of all consequences, hysteria is the most common. In the deficient supply of blood to the brain the faculties of the mind are involved. Confidence of strength gives place to fear, mental vigor to weakness and irresolution. Such indications are as untrue to the real character of the individual, as the physical sensations are false and deceptive.

Such is the nature of the large proportion of cases of persons who come into courts of law for compensation for injuries erroneously deemed to be permanent, bringing with them headaches, spinal pains, tingling of the extremities, impaired vision, loss of memory, and many other symptoms of an unstrung nervous system—a series of grievances of the incurable nature of which an acute lawyer takes care to provide himself with ample testimony, which will always be obtained so long as the diseases of the vascular system and their consequences monopolize a too prominent share of the attention of our profession. I have traced several of these persons in their after-career, the large majority of whom entirely recover. I believe it is to the prevalence of error in the early management of these persons, who are almost invariably subjected to depletive treatment, and to the imperfect knowledge of nervous diseases which prevails in the profession, that large sums are awarded for injuries erroneously supposed to be permanent and incurable. Can it be reasonably expected that the truth will be brought home to the mind of a lawyer so long as our own opinions are yet uninstructed upon it? Sooner or later their true nature will become established facts in the minds of our profession, and we shall no longer hear the painful discrepancies of opinion among medical men that now prevail. The light of improved knowledge will dissolve the mysteries which daily surround these cases in the form of supposed spinal concussions, partial paralysis, effusions into the theca vertebralis, thickening of the membranes of the brain, spinal cord, and lesions of this organ or that. These, as Dr. Sydenham declares, are but imitations and resemblances, and not realities; and that they deceive the multitude is undoubted. When real disease prevails, there is no difference of opinion among medical men as to its existence.

It is a very interesting question to investigate how far the functions of the mind are involved in hysteric disease, and how closely it is connected with it, whether the relation between them is direct and immediate, or remote. In cases of local pain, and also in the local contraction of muscles, arising either from an excessive action of one muscle, or from the loss of harmony of action with its antagonist, as in a permanently flexed forearm or finger, it seems difficult to identify the evil with that part of the brain which we believe to be the seat of mind. And yet an inquiry into the past history of such persons will often reveal the fact that they have been at one time or other the subjects of general or paroxysmal Hysteria, or, in other words, that they have had hysteric fits; and as mental emotion is more or less associated with this form of Hysteria, it would appear not unreasonable to infer some remote relation between the mind and this variety of a disease apparently simply local in its nature. There is something in the mental development of these young persons very characteristic. They are quick and excitable, liable to sudden emotion without adequate cause. In very young persons the local disease may be developed before the mental character is fully matured, but advancing years will exhibit its peculiar features.

It is curious to observe the influence which the nervous system exerts on the daily condition of us all. When unstrung it preys upon ourselves. It is not in the varying force of our pulse, for that gauge is not sufficiently fine to detect the variations of health, that we can refer a consciousness of strength and vigor on one day that fails us on another. It is that our nervous system is more or less relaxed. There is a real illness and a factitious illness, and in this we observe the remarkable influence of mind in exercising a controlling power over the body. People without compulsory occupation, who lead a life of both bodily and mental inactivity—people whose means are sufficiently ample to indulge in, and who can purchase, the luxury of illness, the daily visit of the physician, and, not the least, the sympathy of friends—these real comforts come home to the hearts of those ornamental members of society who are living examples of an intense sensibility, whether morbid or genuine, who can afford to be ill, and will not make the effort to be well. They are, in truth, well or ill, as you choose to take it, and they are only ill because they fail in mental effort, that mental resolution which is sufficiently powerful to rouse the dormant energies of the body and throw off the sensations of lassitude, of unreal fatigue and weariness of body and mind. A poor man cannot afford this indulgence, and so he throws the sensations aside by mental resolution. How often does a sense of weariness and fatigue succumb to active and vigorous muscular exertion?

There is a real fatigue, and a nervous or unreal fatigue. A lady will tell you she was so tired that she could not walk another step. She thinks so, and without an adequate motive she cannot make the required effort. Give her the motive, such as the sudden illness of a relative or friend at a distance, and she will extend her walk to miles without effort or subsequent fatigue. How is this? It is that by a great motive acting through her mind she has called upon those dormant powers of her system which are possessed by all of us to be employed on critical occasions. Rarely, if ever, is the body subject to a degree of fatigue so great that an adequate motive will not obtain renewed exertion. When a lady tells you she can only venture on a walk of half a mile, you will understand that this effort is determined by the ordinary, not the extraordinary motive. It is your duty as her medical attendant to place before her such inducements to a greater effort as shall call on the exercise of her dormant power, *the reserved fund* of physical strength, and she will walk four times the distance without fatigue. A poor man runs a race against time, and reaching the goal he drops from fatigue. Offer him at the moment £100 if he will run one hundred yards farther. He will accept the offer, run the required distance, and then drop. This is resolution acting on his muscular powers through his nervous system, screwed up by an extraordinary mental effort.

And this law of Nature is applicable to us all in our daily intercourse with the world. A man resolves to accomplish a certain amount of work on a given day, and he completes the task he has assigned himself by virtue of his resolution. Such resolution is eminently protective against fatigue.

A question arises to one's mind: is Hysteria what is termed a specific disease, or is it the invariable result of a condition of health into which all persons pass in reduced states of bodily vigor, but only modified in degree? I presume it is associated with a peculiar organism common to man but not involving all, as some persons amenable to the influence of *mesmerism* pass readily into profound sleep, while others are entirely unaffected by it. It notoriously is far more common in women than in men, and in young persons from the age of seventeen to thirty, in the unmarried than in the married. We do not associate hysteric affections with persons of either sex who are characterized by vigor of mind, of strong will, of strength and firmness of character. Such persons may be reduced by protracted illness to a condition of weakness both bodily and mental, but they do not in their reduced strength, so far as I know, exhibit any of the peculiar features of hysteric affections.

There is a remarkable form of Hysteria which affords evidence on this subject. It is notorious that the sight of a person under an hysteric attack has a tendency to involve other hysteric persons around her. It has happened to me several times in my hospital career to witness the contagious, or rather the imitative, form of active or paroxysmal hysteria on a large scale. On one of these occasions, in a ward of twelve females, no less than nine young women were affected at the same time. Several were so violent as to call for the assistance of sisters, nurses, and other servants of the establishment to restrain them; and inasmuch as a person under the influence of Hysteria brings into action all the latent strength of her muscular frame, *which is greatly in excess of her apparent strength*, the services of these attendants were scarcely sufficient for the purpose, several requiring three or four strong men to prevent injury to their persons. The attack commences in the person of one girl who may have been the subject of some trivial operation, or been brought under the immediate influence of the disease by mental emotion. No sooner is the condition of this person observed by her fellow-patients than her influence is felt throughout the ward, and the second subject may become involved, occupying a bed at the remote end of the room, and thus it passes irregularly from bed to bed, each patient appearing to take the disease in the order of their constitutional liability. In the course of an hour, more or less, it subsides, and tranquillity is restored, but the evil only slumbers, and on the following day the same scene may recur: less violent, perhaps, but acted by the same persons as at first. Some of these patients, who were not affected to violence, were affected to tears and wept in silence, while some few were not implicated at all, nor did they show any tendency to sympathize with the disease. These curious attacks, though they appear to the subjects of them irresistible, are yet but the result of what has been termed a "surrender," and might be prevented by an adequate motive. The mode adopted to arrest this curious malady consists in bringing these persons under the influence of some powerful mental emotion, and in making some strong and sudden impression on the mind through the medium of, probably, the most potent of all impressions, fear. They are not lost to consciousness, and for the moment, except in the intensity of their paroxysm, they will listen to the voice of authority. Sympathy and kindness, or tenderness of voice and manner, are worse than useless. They rather aggravate than mitigate the evil. Ridicule, to a woman of sensitive mind, is a powerful weapon, and will achieve something; but there is no emotion equal to fear, and a threat of personal chastisement will not necessarily be required to be carried into execution. On two of the occasions I have referred to, a few quarts of cold water suddenly thrown on the person of a chief delinquent instantly brought the ward to a state of reason and subordination. The disease succumbed to the indignity of the treatment. There can be no doubt, then, that a malady spreading by sympathy and cured by fear, has its origin in the mind. I think you will find on close inquiry that nearly all cases of paroxysmal Hysteria originate in some form of mental excitement, and that of a depressing character, such as sorrow or disappointment. It is not the result of mere emotion. Joy, gladness of heart, or a sense of pleasure, rarely produce it; yet it is difficult to explain either its immediate or proximate origin in attacks occurring during sleep. Sometimes these patients suddenly awake from sleep with severe palpitations of the heart leading on to a direct attack. What can be their immediate cause? There is no disease of the agents of circulation, or any suspected variation in the quantity of blood thrown upon the heart by which to explain it. Whence, then, the eccentric action of this organ? Possibly some mental emotion in the form of a forgotten dream, or some other occult mental operation which escapes cognizance, such as occurs in cases of somnambulism.

In these current remarks on general Hysteria, we must not lose sight of the subject taken in the surgical point of view. I have stated, both in this and in the last lecture, that under the condition of impaired health the nerves of a part of the body may become the subject of a deranged action, by which, as Dr. Sydenham has declared, and we in our generation almost daily observe, so many symptoms of actual disease of that part may appear, as to give the exact aspect or verisimilitude of local organic change of structure when such disease is entirely absent. There may be nothing apparent on a first inquiry to associate the case with Hysteria, whether local or general. It is not necessarily nor commonly preceded by hysteric paroxysms. There may be no appearance of illness, no heat or undue excitement of the system, nothing, in fact, to connect it with hysteric disease; yet it is nothing but local nervous exacerbation, and from the want of a better name we call it Hysteria. You must not confound it with simple neuralgia, and with still less reason, with epilepsy. It is not, however, always easy to draw a distinct line between neuralgia and Hysteria, for both may have a constitutional origin, and be amenable to nearly the same treatment. In neuralgia, we have a more generally local and more persistent affection of a nerve. The disease appears to be limited to the nerve itself, the course of which may be traced by the pain, which is often excessive; whereas in those cases of Hysteria marked by local pain, the pain is general, involving the structures around, in common with real disease of the part affected. In neuralgia the disease is placed on a recognized nerve, and a person is said to have neuralgia of a given nerve, such as the frontal, mental, or digital. In Hysteria any locality may be affected without reference to the distribution of nerves; while epilepsy is characterized by well marked symptoms clearly of a cerebral origin. If you amputate a limb for hysteric pain, you throw the disease back on its nervous centre, and you kill your patient. In the early part of my hospital career I have seen this fact more than once exemplified. In such cases operative surgery is entirely out of place.

Now, before I proceed to illustrate these statements by reference to cases, of which I have a sufficient supply, I wish to make a few remarks on the relation between local Hysteria and the nervous centres, namely, the brain and spinal cord. Any facts that tend to throw even a gleam of light on the connection between them must be interesting.

I refer to the influence of anæsthetic agents, especially of opium and chloroform. In cases of Hysteria marked by local pain, relief is given by the application of opium to the affected part; a fact which does not confirm the generally entertained opinion that the local affection is dependent on irritation of the nervous centre. Select a case of hysteric contraction of the muscles of a joint; the elbow or fingers. If you administer chloroform, the contraction of the muscles which may have existed for months, and which has resisted repeated attempts to extend them, will now yield to a gentle effort of extension, and the limb is immediately restored to apparent repose. Supposing this morbid contraction of one or more muscles to be caused, as we believe, by irritation of the nervous centre, how does opium or chloroform affect it? The effects of chloroform on the circulation are assuredly not in the direction of health, for it converts arterial into venous blood, or, at least, it gives to arterial blood the dark color of venous, and we can hardly believe impaired circulation of a part of the body compatible with its improved function; and yet the disease subsides. This "irritation of the nervous centre," as I told you, does not convey a very clear idea of the nature of the relation between the respective parts, namely, the seat of the disease, and its source or centre. Perhaps the nearest approach we can make to a solution of the difficulty is by saying that these two agents—opium and chloroform—suspend for the time the influence of both sensory and motor nerves, under which suspension the local pain, or the erring muscle, partakes of the general influence of the anæsthetic. To bring this morbid state of the muscle within the influence of the mind as its cause, is almost of necessity to infer the local evil to be wilful; but if it were so, the state of unconsciousness during sleep would remove it, which it does not, for the contraction is constant by night and day, while the specific influence of the chloroform suspends the disease, if it does not cure it, and the renewed, though partial, contraction of the muscle is now prevented by mechanical agency. The remarkable circumstance consists in this, that a disease of long standing, which incapacitates for exercise and occupation, is removed in a few minutes by the agency of chloroform, and the patient placed at once on the high road to recovery. Is this curious fact confirmatory or otherwise of the origin of the disease in the nervous centre?

Unfortunately, hysteric persons have no exemption from real disease, and when the two are found in combination, a difficulty in diagnosis will frequently occur to test the pathological knowledge of the surgeon. The local disease is accompanied by symptoms of an eccentric character that do not legitimately belong to it. Local pains are aggravated in the active stages, and do not subside in a degree proportionate to the local improvement. A small malady, such as a sprained wrist or ankle, is magnified into a large one. The constitutional symptoms take the direction of Hysteria instead of fever. The vascular system indicated by the state of the pulse, the skin, etc., is less involved than the nervous, and months will often expire before recovery is complete. To a surgeon not familiar with hysteric disease, who practises his profession with reference to one only of the two systems of which the body is composed, these cases will always be obscure and difficult of management. When an injury occurs to the person of a young female, and to many others neither young nor female, hysteric symptoms are almost certain to develop themselves in some form or degree before recovery is complete.

FIFTH LECTURE.

HYSTERIA—CASES.

Distinction between nervous and vascular diseases recapitulated—General localities of Hysteria—Case of Hysteria of the muscles of the larynx—Hysteric affection of the mammary gland—Value of exercise—Relative value of foot and horse exercise—General and local treatment—Hysteria of hypochondriac regions—Spinal affections—Cases—Efficacy or inefficacy of issues—Cases—Railway actions and extortions—Cases—Hysteric joints, treatment of—Cases.

BEFORE I quote cases in illustration of the principles inculcated in the last two lectures, so all-important is the subject in the practice of our profession, so grave the consequences of false diagnosis, that I briefly restate the question under our consideration. It is a notorious and undisputed fact that any part of the body can be the seat of such local derangement, either of the nerves of sensation or of motion, as shall accurately represent real disease of that part, when no disease, properly so called, really exists, and that the detection of the truth is only made on a closer observation by the surgeon than is generally awarded to it. The examples of derangement, whether of the nerves of motion or of sensation, are local and general, and are treated both by the physician and the surgeon, but more commonly by the latter, when the evidence of local change of structure (or disorganization) is absent.

As a rule the malady, with the exception of true Neuralgia,

comes under the variety of hysteric disease; and the occurrence of such maladies in the persons of the young, and especially of females whose nervous system is more lightly strung and more readily deranged, clears the path to an early diagnosis. Draw the line, as clearly as you are able, between the diseases of the vascular and the nervous systems. Recall to your minds the local consequences of increased action of the blood-vessels, both immediate and remote: local heat, local swelling, local redness, if the structures involved are visible; and, lastly, local pain. This is inflammation, which may either pass off, leaving the tissues unaltered in structure, or may leave behind it evident marks of its activity in the form of deposits, of thickening, of abscess, or of death. Local nervous affections have little in common with such maladies, and are distinguished from them by the absence of those features which characterize diseases of the vascular system. They have neither heat, nor redness, nor swelling, in a measure, compared with them; and yet the two maladies are daily confounded. Excessive or undue action of the vascular system more readily involves the nervous than the nervous implicates the vascular.

Cases of real Hysteria may be reckoned in multitudes in the practice of any one surgeon. The more common seats are the female breast; the side of the trunk under the ribs; the whole spinal region from the atlas to the sacrum; any joint, but especially the knee; the stomach, the bladder, and the ovaries; the muscular system of the extremities, indicated by spasm or permanent contraction; and the muscles of the larynx. But no part of the frame has exemption from liability, so far as I am aware.

I will first quote a very simple case of Hysteria, the evidence of which is immediate, and the attack transient. In some slight forms the patient loses all command over the voice, which suddenly sinks to an almost inaudible whisper, without any other accompanying symptom. I have seen many examples, but that I give occurred under my own observation, as it has probably in some form occurred under that of others, for it is as old as history. The subject was a young lady of about twenty, as Sir B. Brodie observes, of pale complexion, and having cold hands and feet. While I was engaged in conversation relative to her health, I somewhat imprudently remarked that a mouse was running about under the table at the end of the room. She uttered an exclamation of alarm, and in an instant so entirely lost the power of audible speech that I was obliged to approach her and to put my ear close, to hear her. The ferocious cause of the mischief having paid the penalty of its intrusion by the loss of all it possessed on earth, the lady in the course of an hour recovered her voice. Had this person been in sound and vigorous health, she would probably have sustained the shock to her nervous system with less derangement of it. The case is interesting, as showing the sudden influence of the mind on a particular nerve in the general system. Ammonia, chloric ether, henbane, etc., quickly administered would probably shorten the attack, for which agents brandy is a good substitute.

Cases of hysteric affections of the breast occur in young persons from 16 to 20. They are associated with a disturbed condition of the general system, but not especially with the functions of the uterus itself. Although the catamenia is often deranged and defective, it is not necessarily so. The general system is at fault, indicated by a low circulation; frequently a chlorotic aspect, failing appetite, languor, and indisposition to any form of active bodily or mental effort. In the cases I have seen, the breast has been small and soft. The disease consists of simple pain in the organ, one or both, but more severe in one than the other. It is most active at the catamenial periods. The breast is unaltered in form and substance. The evidence of local inflammation is entirely absent. Leeches, or other form of local depletion, give no relief. Blisters, and irritating plasters and ointments, answer no useful purpose. The degree of pain varies with the condition of the health. For a period it may almost cease; a period coexistent with a change of air and occupation, or a residence of a month or two at the seaside, but the pain relapses on the return of the subject to the ordinary habits of life. The pain is dull and aching, and very unlike the smarting and soothing pain which accompanies abscess or the more formidable diseases.

With this history, who will doubt the constitutional nature of the malady or the efficacy of such remedies as tend to change weakness for strength, to promote appetite, to keep the circulation in action by frequent exercise, taken at least twice daily?

Such exercise should be active; neither strolling nor sauntering out of doors, "to take the air," as ladies term it, nor gardening, nor lounging about; but adopting a good brisk walk, at a pace of at least three miles an hour, *ever stopping short of fatigue.* People will often tell you they "take plenty of exercise about the house, and they are on their legs during many hours of the day." This is fatigue without exercise. What we want for health is exercise without fatigue, for fatigue is exhaustion, and health is to be obtained only on the terms I have mentioned. I do believe there are many maladies, or at least many forms of indisposition, which, with a strong will, may be walked away, provided the exercise be taken systematically and rendered a prominent feature in the daily treatment. The distance walked should be increased daily, and a claim made on increasing strength for increasing exertion. I doubt whether horse exercise, however agreeable or however stimulating both to mind and body, is equal in sanitary value to exercise on foot. In the case of horse exercise the muscular exertion to an experienced rider, male or female, is very slight, and though the distance compassed may be great, the muscular exercise, so far as it is an important element of treatment, falls short of the requirements of health. That the effort is comparatively not great is proved by the long distances ridden and the number of hours during which a delicate girl is seated on her saddle. The general concussion or shaking of the muscular frame incidental to this exercise in an unpractised rider subsides on its frequent repetition, and when the rider becomes familiar with the action of the horse, so slight an effort is requisite to maintain the equipoise of the person in motion, and so entirely do the movements of the rider respond to those of the animal ridden, that the muscular effort amounts to almost nothing. Horse exercise, therefore, cannot strengthen the muscles, because it does not sufficiently exercise them. It is an agreeeable and a useful recreation, but I suspect its influence as a source of health acts more beneficially on the mind than on the body. I do not wish to undervalue exercise on horseback; I do only desire to meet the too general belief that horse exercise can supersede exercise on foot as a means to restore health.

Although you cannot cure these maladies by local treatment, you can generally mitigate the severity of the pain by the application of opium and other sedatives, but I have the greatest reliance on opium as the most efficient. Do not fall into the error of supposing that a "cooling lotion," which shall contain tincture of opium as a component part, can exercise any sedative property on the pain, but resort at once to the extract known under the name of *extractum opii fluidum*. Spread the pure extract on lint, or dilute it slightly with water, and apply it to the breast; give tonic medicines, especially in the form of bark and iron, such as eight or ten grains of the ferro-citrate of quinine, or two drachms of tincture of bark, with a pill of two or three grains of sulphate of iron, twice daily. Tonics are out of place in the after part of the day, and especially after dinner. If the pulse indicates weakness—and it is highly probable it will—give wine frequently, and in sufficient quantity to produce an impression. It is not the consumption of a single glass or more than will effect the end you have in view. You must meet the very marked tolerance of wine by an equally positive increase in the quantity, and adminster it not once, but twice or thrice in the day, *so long as weakness of the pulse and the patient are indicated.*

FIFTH LECTURE. 79

Of examples of hysteric pain situated under the ribs, more commonly on the left side, it is needless to quote individual cases; they are so common. From some cause not very apparent, they are, however, seen less frequently at the present day than formerly. I attended some years ago a young married lady, the mother of three or four children, the daughter of a medical man of large experience, by whose direction she had been cupped about fifteen times over the seat of pain. The malady prevailed in her system in its active form during many years, and she was not free from it when I saw her at the age of 30. This treatment, which included the local application of leeches by the hundred, and blisters the sum of which might be calculated by the square yard, while it gave no permanent relief, has left its mark in more senses than one on the person and constitution of this lady for life. At all events, her excellent parent has the merit of perseverance, if not of discrimination.

Among the cases of the sympathetic or imitative forms of Hysteria which I have already quoted, two of these females exhibited the scars of similar local treatment for supposed organic disease under the ribs, and I have seen many others. I need hardly tell you that this is a constitutional and not a local infirmity, and must be treated accordingly, *or not treated at all.* What structure or organ occupying this region on the left side, under the lower ribs, can be supposed the seat of this pain? It is deep-seated, and therefore the abdominal muscles are beyond suspicion. Is it the colon, or the spleen, or the base of the left lung, or the diaphragm? Whichever structure is involved, if any, rely on it the essence of the malady is seated in the nerves, and in the nerves only. If it were organic disease, its nature would become in time palpable. There is this important distinction between the two affections; that *organic disease has a crisis, and nervous affections for the most part have none.*

Spinal hysteric affections are, perhaps, of all hysteric maladies, of the most common occurrence; happily they are becoming somewhat notorious. It is in the records of pathology, no doubt, that a young female may be the subject of real spinal disease; but where are the cases to be found? You may pass through life and not see two. And while I state this opinion your minds may possibly revert to some case you have already attended which you think exceptional. Look more closely into it, and you will detect your error. In forming an opinion on any given case on which you may be consulted hereafter, you had better make a startingpoint from the knowledge of this fact, that nothing in pathology is more improbable than that a young lady should be the subject of organic disease of the spinal column.

Well! a case presents itself for your opinion. A young female, in any class of life, in apparent health, pale or florid in complexion, bearing in her appearance no indication of disease, complains of pain in her back. This pain may be announced without surgical inquiry, or may be detected only on examination. The spine is exposed while the person is placed in bed. Pressure is made by one or more fingers on the spinous processes of the vertebræ, beginning with the atlas. On reaching perhaps the last dorsal or first or second lumbar vertebræ, the girl utters an exclamation of pain, and she instantly shrinks from the pressure. The examination is renewed again and again with the same result. Twenty-three vertebræ admit of pressure through their spinous processes without causing suffering. Pressure on the particular one, or perhaps two, causes instant and often severe pain. By-the-bye, who ever heard of real disease attacking one or even two vertebræ only?

I have already told you the probable result of this inquiry, but you, I trust, will not be deluded. Be assured to the extent almost of certainty, that there is no organic disease, either of bone or of any other texture.

I attended a girl in St. Bartholomew's Hospital of about twenty years of age. She had the appearance of a strong and healthy person, and there was nothing in her aspect to indicate that she was the subject of disease. Before I reached her bedside the House-Surgeon informed me she was the subject of "spinal disease," and I smiled at his credulity. To the students around I said: "If on examining this girl she makes an exclamation of pain and shrinks from the pressure of my hand, rely on it she has no disease whatever, and that her case is one of simple Hysteria." On reaching the first and second lumbar vertebræ she uttered an expression of severe pain, and nearly threw herself out of bed. The diagnosis was confirmed, and she was treated for a nervous, not a real disease. Extract of opium dissolved in soap liniment was rubbed on the spine for a few days, and then the opium was omitted, and the back generally rubbed by the hand twice daily with some force of pressure. She was ordered valerian, bark, iron, and a full diet, with wine. Her recovery occupied one month.

I was consulted in the year 1862 on the case of a young lady of about 24 years of age. She had had "spinal disease" for several years, and many surgeons of more or less eminence had been consulted on her "very remarkable case." Her aspect was that of a healthy person. She was inclined to be stout, and exhibited no indication of serious disease, or indeed of disease of any kind. During five years her back had been liberally cupped, leeched, blistered, and embrocated without benefit. I was informed that the pain had occasionally intermitted, that her condition had improved for a time and then relapsed, and that although nearly the entire five years had been passed in her chamber and in the horizontal posture, yet that occasionally she would join her family and seek relief from the monotony of her life in the gaiety of the ball-room, where she forgot her diseased spine and all its attendant miseries, and danced for hours with life and animation. I examined her back with more than usual care. The pain, always true to its own locality, occupied the second lumbar vertebræ, and always returned on the pressure of my finger on that particular spot. Occupying her attention by conversation, I gradually subjected the whole back first to gentle, and then to severe, pressure. *With both hands I grasped the trunk, and moved it forcibly in all directions without creating any sensations of pain.* I then passed the flat of my hand rapidly down the spine, employing not pointed, but obtuse pressure over the whole surface, and thus satisfied myself that there was no disease. After the interval of a few minutes, pointed pressure on the second lumbar vertebra produced the same symptoms as at first. On examining the surface I observed the mark of a cicatrix of about three inches in length running along the side of the affected vertebra, and on inquiry I learned that one surgeon whom the family had consulted had deemed it necessary to look within and below the surface, under the supposition that there might possibly be a tumor or some morbid growth, the removal of which would be conducive to her recovery! Nothing, however, was found, and the excision of a small portion, I presume of the erector spinæ muscle, afforded no permanent relief; at least no benefit had arisen from the operation at the expiration of many months, when I was requested to see her. It struck me that this was carrying the experiment of operative surgery rather far, but I did not make any remark to that effect at the time. I certainly made an inquiry as to the product of the operation, and the father of the young lady told me that he was shown something, but he was not competent to state exactly the nature of the substance removed. On discussing the pathology of the case with the family and the attending surgeon, I expressed my conviction of the hysteric nature of the disease, and that the lady was capable of exertion could she be induced to attempt it. I saw at once that I had failed to convey my own convictions to the family, that my opinion was not satisfactory, and that in the judgment of the lady's father, a very sensible person, the opinion of one man could not outweigh that of the many, and that the testimony of the many was the safer guide. The patient returned to her couch, on which she may be now reposing for aught I know to the contrary, for I saw her but once. I had, however, the satisfaction to hear the medical man say as I left the house, "I believe your view of the case is the only true one."

As treatment by means of issues was formerly in great resort, and is yet far from being abandoned as a means of checking the progress of carious disease in the vertebræ, it is worth considering for a moment the principle of their action. To control one disease you make another, which is supposed to act as a drain in carrying off the morbid actions of the original disease by derivation, or *counter-irritation* as it is termed. An issue is an ulcer, secreting matter, and drawing more or less on the powers of the constitution. An ulcer is a disease. All disease exercises a depressing, not an invigorating influence on the system. The sum total, then, is increase, not diminution, of the evil. The morbid condition of true spinal affections is that of *caries* or crumbling of bone, not inflammation. Is it probable that a pair of secreting ulcers can tend to restore bone that is lost? Will the capillaries be more likely to secrete material to be converted into healthy bone within the body, because you have made an ulcer outside? The actions going on within are those of deficiency, not of excess. Here comes in again the old doctrine of inflammation. The operation of an issue is equivalent to that of the lancet, and in these days that instrument has become obsolete in the hands of all sensible and thinking men. I acknowledge with all regret, in looking back at the early part of my own professional career, to have frequently committed this error in treatment, and I willingly make retribution to another generation by declaring my conviction of the entire futility of an issue in this description of disease to answer any useful purpose.

While on a visit to the house of a friend in the country, I was requested to see one of his daughters who had been confined to her room for fifteen months, in consequence, as I was told, of diseased spine. She was twenty-one years of age. Her countenance was pale, but not unhealthy. She had been condemned by a court of surgeons to a long confinement to the horizontal position, and she bore the judgment against her with resignation and humility. From the appliances around her, and the general arrangements of the room, it was obvious that the siege was to be long and vigorously maintained. My visit was not a professional one, and I did not propose to myself at the time to discuss the subject of her illness. Accident brought me in contact with her medical attendant, and in the course of conversation with him some features of her case were mentioned, which appeared not very consistent with real disease of the vertebræ. We examined her carefully, and the consultation which ensued terminated in the proposal that his patient should change the horizontal posture for that of a semi-inclined plane. In a week she sat upright in an easy chair, and within a month she joined the family circle, entered into all their pursuits, and could ride any reasonable distance on horseback without fatigue.

I see no advantage in multiplying these painful examples of hysteric disease. They have all general characters in common, and are amenable to the same principles of treatment. And so with cases of supposed injury from railway accidents. Again and again have I heard medical men, physicians, surgeons, and general practitioners, come into courts of law and state their opinion that the plaintiff had sustained grievous and probably permanent bodily detriment to the spinal column, on the evidence of pain produced by pressure of the finger on one or more of the spinous processes: evidence far more than counterbalanced by the fact that these deluded persons have walked unaided into court, and have stood or sat in the witness-box for three quarters of an hour while under examination.

No evidence of the reputed symptoms of these persons is obtainable either through the eye or the touch of the surgeon. There is nothing palpable, nothing organic. You take the assertions of your patient on trust, you identify yourself with his case; you place an object before him, and he declares he cannot see it; you refer to an occurrence that happened last week, he declares he does not remember it. He suffers incessant pain in his back; he staggers in his walk, occasionally coming to a harmless fall; he has convulsive twitchings in his legs, occurring chiefly in bed, which he says he cannot control. He passes blood in his urine, which always escapes the notice of others, for it is invariably passed at the water-closet and at no other time, but on inquiry you will find that this afflicted person can walk four or five miles; that as regards his vision his iris acts well, and the ophthalmoscope detects nothing; his appetite for food is sufficient for perfect nutrition. And let me ask you finally, whether on these conditions it is more than remotely possible such a person can be the subject of any serious organic disease? But presuming on the possibility of such contradictory evidence occurring in a single and exceptional case of real disease, will your credulity reach so far as to admit of their frequent occurrence? To you such cases will be presented singly and individually, but they are brought into courts of law in multitudes; and with their sympathies enlisted on behalf of the plaintiff, no great skill is required to impress the jury in his favor.

One of such cases I will give you as an example: A man without property or profession brought an action against a railway company for injury to his spine. This statement, on the face of it, is an absurdity. How can a man without property bring an action at law? Well, he applies to a lawyer, who undertakes the case on his behalf, with a certain compact and understanding as to the question of future payment! Thus the lawyer becomes the plaintiff, and the plaintiff the witness in his own case. The man's injury was made out to the entire satisfaction of the jury, and very heavy damages were awarded by them, coupled with severe comments on the negligence of the railway directors.

It was positively known, at the time, by several persons engaged in the action, among whom was a detective officer, that within a few days of the trial the plaintiff, or the witness, whichever you please to term him, *had walked a race* against another man! Yet this man was declared on authority to have sustained a permanent injury of his spinal column!

With respect to hysteric affections of joints, knee cases, etc., they are in truth as common as Sir B. Brodie has declared them to be, and I thoroughly corroborate all he has said on the subject of this most important and interesting disease. Three fourths of all knee cases in the upper classes of society, says this great authority, are not cases of inflammation, though they appear so. There is no organic disease whatever in the joint. They are cases of local pain, originating in impaired health. They are not amenable to treatment for inflammation and its consequences. Your liability to an error in diagnosis is just in proportion to the supposed infrequency of local nervous, as compared with vascular, derangement. The knee is by far the most frequent seat of these affections, and the cases are found among young women not in the lower class of life; but even this class is not exempt. You will find, on the occasion of your first visit, the patient walking lame. This lameness has existed for several days, probably weeks, before attention has been attracted to it, and has come on very gradually. The joint is stiff: not that it won't bend, but the movement is painful. There may be some increased heat in the joint when compared with that of the opposite limb, but not much in degree. The knee is slightly swollen. If you see the case after treatment has commenced, i. e., after the repeated application of leeches, blisters, and tincture of iodine (the almost universal agent in difficulty), the swelling will be palpable, and the outline of the joint has undergone a change. As the case progresses, the lameness increases, but the aspect of the joint remains as in the first stage, neither the swelling nor the heat increasing in the same proportion. In this condition the limb may remain for months, or even for years, subject to the same treatment without improvement. One feature in this case ought to have struck you as worthy of notice, namely, that so many months have passed without organic change; the joint is neither stiffer, larger, nor hotter than it was in the early stage of the treatment. I say it ought to have struck you. Perhaps it has not! The aspect of this lady is that of un-health. She has become pale, partly from depletion, partly from loss of exercise. Her pulse is weak, her appetite bad, and the catamenia, as a rule, defective. You fear to give tonics and alcohol, lest you aggravate the supposed local inflammation.

Having exhausted the negatives in treatment, you now venture on an onward step, and you give bitter infusions, gentian, cascarilla, with ammonia and ether. But you are still behind the necessities of the case; you have adopted from the beginning a false diagnosis, and the difficulty is how to get back to the right groove. There is only one course; begin afresh, and treat your case on a different principle; convince yourself that nerves may go wrong as well as arteries and capillaries, and as you treat excessive action in the blood vessels, rightly or wrongly, by local depletion, so apply such remedies as check excessive action of nerves, in the form of opium, belladonna, chloroform, etc. Build up the health by increasing the force of the circulation. The agents are a thoroughly nutritious diet, wine frequently in small quantities, tincture of bark, iron, fresh sea air, change of locality and associations, agreeable mental occupation. Assure your patient she has no real disease, but the semblance only. Leave the functions of the alimentary canal to take care of themselves. The constipation incidental to a low innutritious diet and an inactive life will subside under the influence of a nutritious one, improved health will restore its function. There is no real harm in a day's constipation; it is sometimes a good. At all events leave the bowels alone. With regard to the joint rub in some blue ointment and extract of opium, in the proportion of one third of the latter, and roll it firmly with a flannel bandage. Encourage moderate daily exercise on a level ground, on a carpet, or on a lawn. If the case is chronic, don't be disappointed if the progress be yet protracted to weeks. The pain and the stiffness may subside very slowly by virtue of their long possession by the joint; but you are in the right path, and rely upon it your patient's recovery will justify the sound principles of your treatment.

In the course of last year I was consulted by the family of a young lady of eighteen years of age, living at a distance from London, relative to an affection of the knee, from which she had been suffering for a period of ten months. The joint was stiff and painful; she moved about on crutches; there was no considerable amount of heat, and what alteration existed in the form and outline of the knee was due to the activity of the past treatment; the tissues had lost their natural softness and flexibility; the joint had been repeatedly leeched and blistered, and subjected to the application of liniments in variety of color and composition; an issue had been made on the inner side of the patella, which, judging from the cicatrix it left behind, had not been a small one, and the curative influence of which had not been discoverable during four months, at the expiration of which nature was allowed to heal it.

I considered this a case of Hysteria on the following evidence. The subject was a young lady of a hysteric age.

She had sustained no sudden injury to the joint, neither blow, nor fall, nor sprain. The malady was gradual and spontaneous. Had the disease been of the inflammatory class, the remedies would have probably long since cured it. There was no appearance of disorganization otherwise than integumental. The pain was generally aggravated at the catamenial periods. Bending the joint afforded no evidence of disease within it, no grating or roughness of the cartilaginous surfaces. The pain varied greatly in intensity at different periods.

This evidence was sufficient, and, to my judgment, conclusive. I strapped up her joint in an opium plaster. She took bark and iron and wine, and in a fortnight began to walk about without her crutches; but two months had elapsed before her recovery.

Many years ago, when I was less familiar with hysteric affections, I attended the case of a young lady of nineteen in conjunction with Mr. Stanley. We both deemed the disease to belong to the class of inflammation, and conjointly adopted the usual remedies so indiscriminately resorted to in all painful affections of joints. Many weeks elapsed without improvement, and I remember that we discussed with some anxiety the probable issue in abscess, destruction of ligaments, absorption of cartilage, and ultimate amputation of the limb!

One day my patient informed me that her sister was going to be married, and that, cost what it might, she had made up her mind to attend the wedding. At this proposal I shuddered. Having expatiated, to no purpose, on the probable consequences of so rash an act, with all the force of language I could command, I determined to give stability to the joint for the occasion, and I strapped it up firmly with adhesive plaster. On the following day I visited her. She told me she had stood throughout the whole ceremony, had joined the party at the breakfast, and had returned home without pain or discomfort in the joint. Within a week her recovery may be said to have been complete.

This case first brought home to my mind the nature and the frequency of hysteric disease.

SIXTH LECTURE.

CASES OF HYSTERIA—*Continued*.

Hysteric affection of the œsophagus—Hysteric affection of the stomach—
—Gastrodynia—of the ovary—Hysteric contraction of muscles—Wry neck
—Contraction of fingers—Contraction of elbow-joint—Hysteric contraction
of the muscles of the leg and foot—Hysteric paraplegia—Hemiplegia.

THE cases of hysteric affections of the knee-joint I have selected for comment in the last lecture may be taken as a type of them all, each case exhibiting some peculiar feature more or less striking on which to fix the attention of the surgeon. It may well be supposed that a large hospital, such as St. Bartholomew's, would furnish examples of them in the course of years in multitudes, and with regard to which I can only refer you to the general description I have previously given of hysteric affections. They may be summed up under the definition of chronic painful affections of joints, attended by stiffness and immobility, without disorganization, and they will only be successfully treated through the agency of the constitution.

I will now give you a few examples of hysteric disease attacking regions less common than those I have yet referred to, but not less distinct in their character:

A young woman, aged twenty-four, was admitted into one

of my wards at the Hospital who was the subject of difficult deglutition. She was a very respectable person in character and position, and had been for several years a much esteemed servant in a good family, and was a young woman of some education. For two months previous to her admission she had complained of difficulty of swallowing her food. As the evil appeared to increase, the family medical attendant was consulted, by whom she was treated for a stricture of the œsophagus. One or more consultations were held on her case, and the œsophagus examined carefully by means of probangs and bougies. These instruments, however, failed to pass a given spot corresponding with the base of the neck, or about one third from the commencement of the tube. She had no local pain whatever. As the obstruction increased, nothing but semi-liquid food passed into her stomach, and this was only effected with a difficult and painful effort. She became emaciated by reason of defective nutrition, and at the time of her admission into the Hospital was weak and somewhat attenuated in form. For many weeks she had taken no description of solid food, and even liquids passed the obstruction with difficulty. The malady now assumed a serious form, and with a view to additional advice she was sent to the Hospital. The case was reported to me on her arrival as that of "stricture of the œsophagus," and I will tell you the preliminary train of thought that passed rapidly through my mind before I opened my lips to the students on the subject. (1.) Real stricture of the œsophagus is at all times a rare disease. When present it is almost invariably a cancerous affection. Cancer is a very rare disease at this young woman's age. For so serious a malady as cancer she does not look ill enough; for though the presence of cancer of the breast occurring at a later period of life by twenty years may be for a time compatible with fair average health, cancer of the œsophagus stamps the constitution early. (2.) She is of an hysteric age, and though thin, she does not look absolutely ill. There is nothing of disease in her aspect, nothing that may not be referred simply to defective nutrition. The history of her case was given me by her medical attendant, who was present on the occasion, and I had no hesitation in recording her disease as that of " Hysteria." I declined the use of a probang or bougie which lay on the table before me, and I simply said, "We will endeavor to remove the obstruction without the aid of instruments of any kind." Her catamenial discharge had been regular throughout. I ordered her bark, iron, valerian, wine, milk with brandy, each to be given in the largest quantities at the shortest intervals *consistent with reason and moderation;* three times in twenty-four hours, enamata of thick soup with an ounce of brandy. These various agents were absorbed into her system with the greatest advantage to her health. Within a week she could swallow finely minced animal food, and in three weeks she ate a portion of rump-steak without difficulty, and was, in fact, convalescent. She was in high spirits at her recovery, and the only vexation she suffered arose from my refusal to pass a probang down her throat before she left the hospital. This I peremptorily declined to do, assuring her that a probang of rump-steak was a far more efficient test of her recovery than any instrument in surgery bearing that name.

A young lady of eighteen, and of slight form, was brought to me from the country with *gastrodynia*. For upward of a year she had suffered intolerable pain in the stomach on taking food of any description. She was much emaciated, and her pulse extremely feeble. Neither trouble nor expense had been spared in her treatment. Her family had consulted medical men of eminence in more than one metropolis, but the severity of the pain continued in spite of treatment. On entering the drawing-room, I heard the sound of suffering from an adjoining room, and I was told that my future patient was paying the penalty of a slight meal of arrow-root, of which she had swallowed a few table-spoonfuls only. Having intruded myself into the room somewhat unexpectedly by its occupants, I saw this young lady in a condition of great suffering, in the upright position, leaning her head on her mother's shoulder, and sobbing painfully. In the course of a quarter of an hour I had obtained some insight into her case, but I could not fail to observe that the mother habitually interposed replies to questions addressed to the daughter, and I explained to her the necessity of my obtaining the answers to my inquiries direct from her daughter. At my request she left the room. Up to that time I had but an imperfect knowledge of the case, but I then led the conversation to subjects which carried the girl away from her malady and all its associations. I spoke of her home and the scenery around it, of which I described the general characters, and enlarged on the beauty of the neighborhood, the lovely rides and excursions, etc., and in all of which I was tolerably successful, considering that at that time I had never seen it. However, the description was sufficiently accurate for my purpose, for it succeeded in distracting the young lady's attention from her suffering, and during the few minutes which this conversation occupied she was to all appearance entirely free from pain. She talked freely and cheerfully, and not the slightest reference was made by either of us to her former suffering. I then changed the subject by saying, " I think your pain has flown away," when she immediately resumed her crying fit, and sobbed as before. She assured me she was in great pain, and that the sensation had been but suppressed. That this was a case of severe Hysteria was highly probable even had I gone no further with the evidence, relative to which the following thoughts occurred to my mind. What could be the nature of this pain if not hysteric? I was told by her family that pain followed the act of deglutition, not remotely or at an interval of one or more hours, but almost as immediately as the food could reach the stomach. This could not be dyspeptic or common gastrodynia, which waits on the process of digestion, and rarely occurs within a period of two hours of taking food. There is one disease only of the stomach in which pain follows the admission of food into it, and that occasionally only, namely, cancer. Was it probable, or scarcely more than possible, that this girl of eighteen could have been for so many months the subject of undetected cancer of the stomach? If cancer, could the attendant pain, so severe as it appeared at the commencement of my visit, be suspended by conversation? It was neither gastrodynia, the result of indigestion, nor cancer; and if not, what remains behind to elucidate the case? It could be nothing but Hysteria, and Hysteria alone could solve the mystery. But she had been treated for gastrodynia and treated for cancer, but she had not been under treatment for Hysteria, simply because these varieties of local Hysteria have never yet fixed themselves on the attention of the profession. To tell a practitioner of the old school that a young lady was the subject of Hysteria of the stomach would be to raise a smile at your expense.

It would be an unprofitable employment of our time were I to enlarge on the subject of the previous treatment. The remedies included, in different proportions and in varying doses, ammomia and other alkalies, under the mistaken supposition of acid secretions; opium in various forms, creasote, bismuth in small and large quantities, mineral acids, etc.

How difficult it is to ascertain beyond all question the real value of many drugs in daily use among us! Although, in common with others, I have frequently employed the trisnitrate of bismuth, I have to this hour no conviction of its utility.

This case did not terminate so satisfactorily as I hoped. That it was a case of Hysteria admits of no doubt, but I had difficulties to contend with in the domestic management of the young lady. Although her symptoms remitted greatly under the use of remedies, she did not entirely recover in the brief period of three or four weeks during which she remained under my care. I gave her small doses of ferrocitrate of quinine, two or three times a day a wine-glass of port wine boiled with spice, and I ordered a plaster of the fluid extract of opium to be applied on the *epigastrium*. If, coupled with these remedies, I could have separated the girl from her family, whose sympathies with her were far too redundant for her benefit, I think she might have been cured in one month. In a case of this kind good domestic moral treatment is indispensable to success.

Hysteric affections of the ovaries are extremely common. Several of such examples I attended with my late friend, Dr. Rigby. The cases I have seen have occurred in young females of between twenty and twenty-five years of age. They are characterized by deep-seated aching pain in the region of the ovary, about two inches above the crural arch. My own observation would lead me to say that the right organ is more frequently affected than the left, but this is probably accidental. Like other hysteric affections, its severity varies with the constitutional health, mental and bodily. It yields but slowly to remedies, and though mitigated, it often returns at longer or shorter intervals. I have applied opium locally with advantage, but an entire change of air, scene, and occupation, combined with tonic treatment, are indispensable to recovery.

There is no class of hysteric affections more interesting to the surgeon or more critical than cases involving morbid contractions of a part of the muscular system. These affections of the spinal nerves are to be found in abundance in all our public hospitals, where in times not long past, if entirely past yet, the knife has been too often called into requisition to settle the question of an obscure diagnosis, to the damage of the patient and the discredit of the surgeon.

Muscles were cut asunder and limbs were amputated, and the disease yet remained behind, to develop itself in some shape yet more formidable, and now beyond the range of cure. On such cases, several of which are in my memory, I shall not enlarge. I prefer to look forward, and to discharge the duty I have undertaken by teaching you the best means of avoiding their recurrence.

A young woman, aged about 27, the daughter of a small tradesman, was the subject of contraction of the *sterno-mastoid* muscle on the left side of the neck, by which her head was much drawn to the side, and her personal appearance disfigured. The surgeon in attendance upon her proposed to divide the muscle at its tendinous origin from the sternum; but before the operation I was requested to see her. She was of a nervous temperament, and was out of health at the time. I considered the disease to be Hysteria, and nothing beyond it. The fixed contraction of the muscle had occurred without local injury of any kind, and it had increased gradually to its then condition. I recommended such general medical treatment as we find useful in similar cases of Hysteria. Failing to influence the opinion of the attendant surgeon, I suggested a second consultation, mentioning the names of one or two surgeons of experience in such cases, who, having visited the patient, expressed a similar opinion to my own.

The operation was performed, and though I was not present I will guarantee the entire division of the muscle as proposed; but it was not attended with benefit, and a second attempt to remove the deformity was undertaken with a similar result. You will not unreasonably infer that the operator in this case—a man of great knowledge and large experience—was not familiar with this class of disease; and he had the candor to acknowledge it. Need you have more conclusive evidence that we have not yet acquired all the requisite knowledge on the subject of local Hysteria?

A young lady of 17, high born, surrounded with all the comforts which wealth and parental care could furnish her, became the subject of contraction of the fingers of the right hand, which was entirely closed in flexion. Any attempt to open the hand was attended by severe pain. She was of a nervous temperament, but failed in no respect in mental power or moral force of character. Having been for some months under treatment in England, she was sent to Paris with a view both to education and to surgical treatment. At the expiration of six months her hand had undergone no improvement, and she returned to London, when I saw her. I should have supposed there could hardly be two varying opinions as to the nature of this morbid action of the flexor muscles, and that nothing but Hysteria could explain a condition of the muscular system which existed in a young girl of 17, who had sustained no local injury to account for it. In this, as in many other similar cases, there was no catamenial derangement. I ordered her a full nutritious diet, exercise, and tonic remedies in the form of bark, iron, and valerian, and she rerurned to her home and family in the country. Some weeks afterward, finding the condition of the hand not improved, she returned to London, and I put her under the influence of chloroform, and the flexor muscles immediately relaxed.

I opened the hand without the smallest effort, and applied a well padded splint to the front of the forearm, extending to the extremity of the fingers. The splint was retained for some weeks, and occasionally removed for a time and replaced. The muscles recovered their healthy tone, and the hand its functions, with the exception of the little finger, the movement of which still continues, after an interval of three years, in some measure restricted.

A young lady, aged 18, of nervous and susceptible temperament, was brought to me by her mother with an affection of the left elbow-joint. The joint was fixed in half-flexion and was supported by a well made leather splint, and had been carefully bound with straps and rollers. I learned that she had sustained no local injury to the arm, but that it had been treated for some weeks for inflammation of the elbow joint; but as the disease did not yield to the treatment, I was consulted. Further inquiry into the history of the case led me to the belief that the affection was of a hysteric nature. I removed the splint, not without some expression of alarm on the part of the mother as to the consequences. The joint itself was neither swollen, hot, nor painful, but any attempt to straighten or move it caused great pain. I explained my views of the disease to the young lady's mother, and recommended a certain course of treatment, which the lady, somewhat, as I thought, against her own judgment and inclination, promised to follow; and I was brought into communication with the medical attendant of the family in the country, to whom I explained my views, and to whose judicious charge I consigned her. Up to this time, and beyond it, I believe my opinions were not shared by either the family surgeon, by the mother, or by the daughter, and I stood alone opposed in opinion to more than one eminent authority in our profession. On the occasion of her first visit, on removing the splint, I placed the arm in a light sling. I advised the free application of opium, with some blue ointment, to the arm, about and around the elbow-joint, which I directed to be firmly rolled; and I prescribed iron, bark, and wine, and advised the lady to be taken into the country. Some weeks elapsed without material improvement, when the young lady was again brought to London for the purpose of consultation. To this end, four surgeons of well-known eminence were assembled. Now observe, the largely prevailing opinion at that consultation was, that the elbow-joint was the seat of disease—that is, of inflammation—and that the inability to move it was due to disease of the joint, and nothing less. I pointed out the absence of local pain, the normal appearance of the joint, the hysteric character of the patient, and I ascertained beyond doubt the fact that she had not long prior been the subject of a true hysteric paroxysm. Before I left the house my views were adopted by the majority, and on the following day, while under the influence of chloroform, I straightened her arm without effort or difficulty, and fixed it in this position by means of a splint. Several weeks elapsed before she regained the entire command over the muscles of her arm, but her recovery was complete.

When this morbid contraction of muscles has been once struck down, the muscles fully lengthened out, and the position retained by a mechanical impediment to further contraction, of force sufficient to control the evil, the morbid tendency yields, and the employment of the splint for a week or two may suffice; but the disease may stealthily return, and the limb should be still kept under some form of restraint for several weeks.

In the year 1864 a young lady, of sixteen years of age, was placed under my care under the following circumstances: For eight months prior to her visit to me she had been suffering from inversion of the left foot, which was so twisted as to bring the point of the foot to the opposite ankle; in fact, at nearly a right angle with the foot of the opposite side. Her family consulted a surgeon of much experience in the treatment of distortions, and of orthopædic notoriety. The case was considered as an example of an ordinary distortion, and the foot was placed in a very elaborately made foot-splint, by the force of which it was made to approach a parallel relation to the opposite foot; but it was an approach only, for no mechanism could retain it in a perfect position, the toes yet in some degree pointing inward. Months elapsed, and the disease continued unchanged. A second orthopædic authority was then consulted in conjunction with the first, and, as no new light was thrown on the disease by the combined opinions of the two, the same principle of treatment was recommended to be continued, and the mechanism was yet somewhat more elaborated. And thus the eight months of the young lady's life passed away, during which no constitutional treatment of any kind was resorted to, and loss of exercise—for she walked, it is almost unnecessary to say, with great difficulty—with other attendant evils, exercised a great prejudicial influence on her health.

When the apparatus, which she had so long worn, was removed on the occasion of her visit to me, the foot immediately resumed its twisted form. The appearance of the limb was singular. Its attitude was that of complete inversion. I have studied the action of muscles a good deal, but I should have found it difficult to explain by what muscular agency of the foot this position was obtained, so great was the inversion. One fact was quite obvious, that it could not be due to the muscles of the foot only, but that those of the whole limb tending to inversion must be more or less involved. The very attitude of the limb was an abnormal one. The disease had appeared almost suddenly in a person hitherto healthy up to fifteen years of age. It could not be due to congenital deformity, and the limb gave no indication of disease or disorganization. There was neither heat, nor pain, nor swelling. In this case, also, there was no catamenial derangement.

I removed the apparatus from the foot, bandaged the limb with a calico roller, ordered a full nutritious diet, with bark and iron, and, having explained the nature of the disease to the friend, sent the young lady home into the country, recommending her to rely on the kindly offices of nature; the greatest of all doctors, orthopædists not excepted. At the end of a month some progress had been made, but not a great deal. She still walked with much difficulty, but it was obvious that she was improving in health and vigor of system. At the expiration of six weeks she accompanied her family to a ball, her foot, as she entered the ball-room, being not yet restored to its normal position. She was invited to dance, and under this novel excitement she stood up, and, to the astonishment of her family, she danced the whole evening, having almost suddenly recovered the healthy muscular actions of the limb! She came to see me two days afterward. She walked perfectly well into my room, and paced the room backward and forward with great delight. The actions of the limb were thoroughly restored, and all trace of the previous malady had disappeared.

These cases, which you will meet with hereafter, show that under a morbid condition of the nerves or nervous system of a given part of the body the muscles supplied by them may become the subject of hysteric contraction or spasm, which can only be treated effectually through the constitution, whether the disease has sprung from the nervous centre or has originated in the nerve itself, either in its course or in its distribution in the muscle. Where a single muscle is involved, or, still more, a portion of a muscle, as in the case of permanent contraction of one finger, one can scarcely explain the fact on the supposition that the disease extends so far back as the spinal cord, and which can implicate one filament of the nerve only and leave the rest unaffected. Still the probability of its seat in the cord is strengthened by the evidence of cases in which the nervous system of a considerable part of the body is involved, as in cases of hemiplegia or general paralysis, and of such I have seen several examples. Their true diagnosis will only be obtained on the most careful investigation. Hysteric paraplegia may sound strange in your ears, but if you will study hysteric diseases you will find no difficulty in the conclusion that Hysteria, as Dr. Sydenham says, may attack any part of the body, and accurately represent the diseases of that part.

A young woman, aged twenty, was admitted into St. Bartholomew's Hospital with the total loss of motor power of the right arm and leg, and partial loss of sensibility. She had been the occupant for two years of a country workhouse, where she was placed on the list of incurables. She had been treated actively for hemiplegia, with cupping and blistering and purgation, and no doubt with a rigid "attention to diet!" Notwithstanding these antecedents, the girl was florid, and presented the aspect of a moderately well-nourished person. This feature in her case struck me as remarkable, namely, that a girl of twenty, who had been confined to her bed in a country union for two years, with so serious a disease as hemiplegia, should present so healthy an appearance, and so unlike what might reasonably be anticipated from the presence of a protracted disease. Then, again, we do not find examples of this form of general paralysis in young persons. Such cases must be very exceptional, and I could not remember ever to have seen one, and least of all should I expect to find such a disease in a hard-working country girl. I observed that the motor power of both hand and foot was totally lost. Volition did not reach the extremities in any observable degree, and in the cases of paralysis I had seen, some power of motion, however slight, remained in the toes, but here there was none. The leg lay perfectly dead to all motive power.

With this evidence the case was recorded as "hysteric hemiplegia," and as such I treated it. The girl was ordered animal food twice daily, bark and iron, and wine in small doses, gradually advancing to larger. The affected limbs were rubbed for a quarter of an hour twice daily. Within a week slight motion and increased sensibility had returned in the fingers, and shortly afterward in the foot; within three weeks she sat up in bed. In the ensuing fortnight she walked about the ward with the aid of crutches. At the expiration of six weeks she returned convalescent into the country, and for many months enjoyed fair average health; but I believe her entire convalescence was temporary only.

There is something more than mere shades of difference between a large metropolitan hospital and a country workhouse; between a school for medical instruction and an asylum for paupers; between an institution which commands the large experience of the many, and one which depends on the limited authority of an individual surgeon, and *consequently* the recovery of this girl was incomplete; but the character of her disease was, in my opinion, thoroughly established in St. Bartholomew's Hospital.

A young lady, aged twenty-two, sustained a shock in a railroad accident in the year 1863. There was no evidence that she had been actually struck, and although she was thrown down with some force she was not incapacitated, because she not only got out of the carriage unaided, but she was able to give some assistance to others who were really hurt. At the expiration of a week she was under the charge of the family medical attendant with the loss of motor power of the lower extremities. The power of sensation was somewhat impaired, but not considerably. Some months elapsed without improvement. She was treated in the mean time for concussion of the spine, followed by supposed effusion into the theca, and chronic inflammation of the membranes of the cord!

I wish I could entertain the faintest hope that my diagnosis powers could ever attain this amount of precision, but it is not a very unusual opinion in cases of this character expressed by medical men.

These consequences of the injury were considered permanent, and an action was brought against the Company for large damages. At this period I saw the case and expressed an opinion as to the probable hysteric nature of the affection, and I advised a compromise with the Company, who awarded the lady a considerable sum in compensation for the real injury she had sustained. Change of air and scene, mental occupation, tonic treatment, adopted in despite of effusion and thickening of the membranes of the cord, but which of the three membranes was never positively stated, was followed by slow, but steady improvement, and in the course of three months the young lady was able to walk about without support, and I subsequently heard that her recovery was complete.

A young woman sustained a railway shock, but, as in the last case, not immediately on the occurrence of the injury. She had total loss of both sensation and motion of the whole of the left side of the body, and partial loss of both on the right. Some months elapsed before I saw her. She had had extreme tenderness throughout over the upper lumbar vertebræ. The loss of sensibility was so complete that she was quite unconscious of the puncture of a needle. The paralysis was referred to local lesion or disorganization of the cord and its membranes. It was difficult to explain the severe lumbar pain at the expiration of months on the supposition of real injury, the very severity of which pointed to hysteria, and there were no circumstances in the case, in my opinion, inconsistent with that diagnosis, and for which malady she was treated.

Her recovery, which occupied many months, was, however, complete.

I have records of other cases of the same character, which I have no time now to detail.

I pray you to recollect, and to keep always before your minds in your future practice, that the human body may be the subject of a class of diseases involving the nerves or nervous system, whether of a small or a large portion of the frame, which are essentially and pathologically distinct from diseases of the vascular system; that they carry with them so much of a resemblance to real disease or disorganization of structure as to deceive the most experienced, but that they are destitute of some of the most prominent features which characterize real disease; that the varieties of this affection, whether in situation, in form, or in intensity, are great; that the source of the affection is not a local one; that, as real disease very commonly holds a relation with the functions of the heart and arterial system, that of local or general hysteria is the product of a disturbed, not a diseased, condition of the brain or spinal cord; and that, finally, in considering the entire phenomena of hysteric affections, it is difficult to deny their relation to the mind, which appears to exercise some mysterious or occult influence over them.

THE END.

ADVERTISEMENTS.

NOW IN PRESS,

AND WILL BE ISSUED EARLY IN MAY,

MAN:

WHERE, WHENCE, AND WHITHER,

Being a glance at Man in his Natural History Relations.

BY

DAVID PAGE, LL.D., F.R.S.E., F.G.S.,

AUTHOR OF "PAST AND PRESENT LIFE OF THE GLOBE," "PHILOSOPHY OF GEOLOGY," "GEOLOGY FOR GENERAL READERS," ETC., ETC.

Tinted Paper.   12mo.   200 Pages.   Extra Cloth.   Beveled Edges.

Price $1.75.

Sent free by mail on receipt of price.

This very remarkable book is one which is destined to exert a striking influence on the current of human thought, relative to the Natural History of Man. As bold as Darwin, and treating of a kindred subject, Dr. Page is even more interesting, because he writes upon a topic which more intimately concerns the human race.

MOORHEAD, SIMPSON & BOND,
PUBLISHERS AND PRINTERS,
60 Duane Street, New York.

ADVERTISEMENTS.

Messrs. MOORHEAD, SIMPSON & BOND will issue in a few days

"BIANCA CAPELLO;" A TRAGEDY.

By LAUGHTON OSBORN. 12mo. 212 Pages. Uncut. Beveled Edge. Extra Cloth. Price $1.75.

BY THE SAME AUTHOR, UNIFORM WITH BIANCA CAPELLO:

"CALVARY" AND "VIRGINIA": TRAGEDIES.

One Volume. Price $1.50.

"THE SILVER HEAD"; "THE DOUBLE DECEIT": COMEDIES.

One elegant Volume, with gilt side-plate from the Antique. Price $2.

"ALICE, OR THE PAINTER'S STORY"; A METRICAL ROMANCE.

One Volume. Price $1.50.

Messrs. MOORHEAD, SIMPSON & BOND will have ready in a few weeks, a new issue of "CALVARY" and "VIRGINIA," $1.50; also, the same bound up with "BIANCA CAPELLO," with collective title, as forming the first volume of Mr. OSBORN's dramatic writings. Fine Paper. Uncut. Morocco Cloth. Beveled Edge. Price $2.75.

PREPARING FOR PUBLICATION,

"THE MONTANINI" AND "THE SCHOOL FOR CRITICS": COMEDIES.

By LAUGHTON OSBORN: being in continuation and completion of the fourth volume of the same series.

MOORHEAD, SIMPSON & BOND, PUBLISHERS, NEW YORK.

A TREATISE ON
EMOTIONAL DISORDERS OF THE SYMPA-
THETIC SYSTEM OF NERVES. By WM. MURRAY, M.D., etc., Physician to the Dispensary, and to the Hospital for Sick Children, and Lecturer on Physiology in the College of Medicine, Newcastle-on-Tyne. Second American edition. Cloth, $1.50. Sent free by mail on receipt of price.

FELIX VON NIEMEYER'S CLINICAL LEC-
TURES ON PULMONARY PHTHISIS. Translated by permission of the Author from the second German Edition. By J. L. PARKE. Sent free by mail on receipt of price. Cloth, $1.50. Paper, $1.25.

HYSTERIA.—REMOTE CAUSES OF DISEASE IN GEN-
ERAL. TREATMENT OF DISEASE BY TONIC AGENCY. LOCAL OR SURGICAL FORMS OF HYSTERIA, etc. Six Lectures delivered to the Students of St. Bartholomew's Hospital, 1866. By F. C. SKEY, F.R.S.
The well-known ability and experience of the distinguished author of this work will insure for it the respectful consideration of the Medical Profession. It is truly admirable, both in form and substance, and will doubtless accomplish much good through those who follow its progressive and scientific teachings. Sent free by mail on receipt of price, $1.50.

LIGHT: ITS INFLUENCE ON LIFE AND
HEALTH. By FORBES WINSLOW, M.D., etc. This important work is written in the most simple and perspicuous manner, and is one of those contributions to science which all can understand and appreciate. It is, indeed, a charming treatise on a subject of vital interest. Sent free by mail on receipt of price. Cloth, $1.75.

ON CHRONIC ALCOHOLIC INTOXICATION.
With an inquiry into the influence of the Abuse of Alcohol, as a predisposing Cause of Disease. By W. MARCET, M.D., F.R.S., Fellow of the Royal College of Physicians, etc., etc. In convenient size, large type. Sent free by mail on receipt of price. Cloth, $1.75.

PATHOLOGICAL ANATOMY OF THE FE-
MALE SEXUAL ORGANS. By Prof. JULIUS KLOB, of Vienna. Translated by Drs. J. KAMMERER and B. F. DAWSON. First volume, containing the Diseases of the Uterus. Cloth, $3.50. Sent free by mail on receipt of price.

THE PRINCIPLES AND PRACTICE OF
LARYNGOSCOPY AND RHINOSCOPY in Diseases of the Throat and Nasal Passages. Designed for the use of Physicians and Students. With 59 Engravings on Wood. By ANTOINE RUPPANER, M.D., M.A., member of the American Medical Association; of the Massachusetts Medical Society; of the County Medical Society of New York, etc.
Sent free by mail on receipt of price. $2.00.

MOORHEAD, SIMPSON & BOND, PUBLISHERS, NEW YORK.

THE DARTROUS DIATHESIS, OR ECZEMA

AND ITS ALLIED AFFECTIONS. By A. HARDY, M.D., Physician to the St. Louis Hospital, Paris. Translated by HENRY G. PIFFARD, M.D., late House Surgeon Bellevue Hospital; Physician to the Skin Department, Out-Door Bureau, Bellevue Hospital; Assistant to Chair of Theory and Practice, Bellevue Hospital Medical College. Sent free by mail on receipt of price. Cloth, $1.25. Paper, $1.00.

FATHER TOM AND THE POPE; OR, A NIGHT

IN THE VATICAN. With a Preface and Ante-Preface by FRED'K S. COZZENS. Cloth, 75c. Sent free by mail on receipt of price.

SLAVE SONGS OF THE UNITED STATES.

Set to Music. A new edition of this popular work is just issued. Cloth, $1.50. Sent free by mail on receipt of price.

The Press says of it:

"We welcome the volume before us—the first collection of negro songs, words and music, that has been made."—*N. Y. Independent.*

"Possesses a curious interest for the student of African character."—*N. Y. Tribune.*

"These endemical lays are, in fact, chief among the signs and evidences of the normal African character."—*N. Y. World.*

THE GLAD NEW YEAR, AND OTHER POEMS.

By ETHEL WOLF. A very attractive volume. Sent free by mail on receipt of price. Cloth, $1.25.

LIFE AMONG THE MORMONS; A JOURNEY

TO THE CITY OF ZION. By A U. S. OFFICER NOW LIVING IN UTAH. One of the most interesting and readable books ever issued on polygamy, and the doings of the Saints.
Cloth, $1.25. Sent free by mail on receipt of price.

WILL BE ISSUED MAY 1, 1868.

THE AMERICAN JOURNAL OF OBSTETRICS

AND DISEASES OF WOMEN AND CHILDREN. Edited by E. NOEGGERATH, M.D., Physician to the German Hospital and Dispensary, and B. F. DAWSON, M.D., Assistant to Professor of Obstetrics in the College of Physicians and Surgeons.

The First Number will be published on the first of May next, and quarterly thereafter. It will consist of 96 pages, elegantly printed, and will contain Original Articles, Reports of Societies, Hospitals, Lectures, and a complete review of Foreign and Domestic Literature of the above subjects.

Articles for the Journal, and subscriptions, are solicited, which, together with all communications, are to be addressed to the Publishers. Terms, $3.00 a year in advance, single copies, $1.00.

MOORHEAD, SIMPSON & BOND, PUBLISHERS,
No. 60 DUANE STREET, NEW YORK.

www.ingramcontent.com/pod-product-compliance
Lightning Source LLC
Chambersburg PA
CBHW020136170426
43199CB00010B/763